Data Mining in Smart Grids

Data Mining in Smart Grids

Editor

Alfredo Vaccaro

MDPI • Basel • Beijing • Wuhan • Barcelona • Belgrade • Manchester • Tokyo • Cluj • Tianjin

Editor
Alfredo Vaccaro
University of Sannio
Italy

Editorial Office
MDPI
St. Alban-Anlage 66
4052 Basel, Switzerland

This is a reprint of articles from the Special Issue published online in the open access journal *Energies* (ISSN 1996-1073) (available at: https://www.mdpi.com/journal/energies/special_issues/Data_Mining_Grid#).

For citation purposes, cite each article independently as indicated on the article page online and as indicated below:

LastName, A.A.; LastName, B.B.; LastName, C.C. Article Title. *Journal Name* **Year**, *Article Number*, Page Range.

ISBN 978-3-03943-326-1 (Hbk)
ISBN 978-3-03943-327-8 (PDF)

Contents

About the Editor

Alfredo Vaccaro received his MSc. degree cum laude and commendation in Electronic Engineering from the University of Salerno, and his Ph.D. in Electrical and Computer Engineering from the University of Waterloo, Ontario, Canada. From March 2002 to October 2014 he was an Assistant Professor of Electric Power Systems at the Department of Engineering, Faculty of Engineering of the University of Sannio. Since November 2014, he has been an Associate Professor of Electric Power Systems at the Department of Engineering of the University of Sannio. He is the Editor-in-Chief of Technology and Economics of Smart Grids and Sustainable Energy – Ed. Springer Nature. He is an editor of the *IEEE Transactions on Power Systems* and *IEEE Transactions on Smart Grids*. He is the Chair of the Task Force "Enabling Paradigms for High-Performance Computing in Wide Area Monitoring Protective and Control Systems" of the IEEE PSOPE Technologies & Innovation Subcommittee. He is the Chair of the IEEE PES Awards and Recognition Committee. He has published more than 140 papers in international journals and conferences.

Preface to "Data Mining in Smart Grids"

Data-driven techniques have been recognized as the most promising enabling technologies for improving decision-making processes in smart grids, providing the right information at the right moment to the right decision-maker. In this context, grid sensor data-streaming cannot provide the smart grids operators with the necessary information to act on in the time frames necessary to minimize the impact of the disturbances. Even if there are fast models that can convert the data into information, the smart grid operator must deal with the challenge of not having a full understanding of the context of the information, and, therefore, the information content cannot be used with any high degree of confidence.

To face these challenging issues the development of advanced forecasting models represents an important issue to address, since they can support the conceptualization of proactive control systems, mitigating the impacts of critical contingencies. To face this problem, in paper [1] a novel load forecasting method, which is based on the combination of different forecasting models and time-series clustering has been proposed. The different models, which include trigonometrical transformation, Box–Cox transformation, autoregressive moving average (ARMA) errors, trend and seasonal components, double seasonal Holt–Winters, fractional autoregressive integrated moving average, ARIMA with regression, and neural network nonlinear autoregressive, are amalgamated on the basis of both normalized periodogram-based distances and autocorrelation-based distances. The obtained results demonstrate the effectiveness of this novel approach compared to other traditional forecasting algorithms proposed in the smart grid literature.

Forecasting of renewable power generation represents another relevant issue to address in modern smart grids, in order to mitigate the impacts induced by the randomness and not-programmable operation of these generating units. This complex process requires massive data processing, especially the analysis of meteorological data obtained by numerical weather prediction models (NWP). To address this critical issue, in paper [2] a new algorithm for effective NWP data preprocessing, which is based on t-distributed stochastic neighbor embedding, is proposed for both reducing the data volume and improving the correlations of wind farm operation predictions. The proposed algorithm normalizes the data collected in order to mitigate the influence induced by different dimensions, and reduces the dimensionality of the NWP data related to wind farm operation. The obtained results show the effectiveness of this algorithm compared to a traditional solution technique based on principal component analysis. Moreover, the dimension reduction preprocessing also had a visual effect, which could be applied to big data visualization platforms.

The large data streaming generated by smart grid sensors, if properly processed by advanced computing paradigms, can enable the development of advanced functions aimed at improving the condition monitoring of power equipment. In this context, paper [3] proposes case-based reasoning to detect partial discharge in substations. The main idea is to estimate the correlation degree between the sensed data and the historical information, in order to detect possible partial discharge using a matching method based on a variational autoencoder network. To verify the effectiveness of the proposed method, a real dataset of historical observations was established through a partial discharge experiment and live detections on the substation site. The obtained results demonstrate the effectiveness of the proposed method compared to other traditional feature extraction methods, which include statistical features, deep belief networks, deep convolutional neural networks, Euclidean distances, and correlation coefficients.

Data-driven techniques can play an important role in improving power system reliability, by allowing the correct system operation also in the presence of severe internal and external disturbances. Amongst the possible phenomena perturbing correct smart grid operation, the predictive assessment of the impacts induced by extreme weather events has been considered as one of the most critical issues to address, since they can induce multiple, and large-scale system contingencies. In this context, paper [4] proposes two methodologies, which are based on Time Varying Markov Chain and Dynamic Bayesian Network, for assessing the power system resilience against extreme wind gusts. Several case studies and benchmark comparisons demonstrate the effectiveness of these methods in the task of assessing the power system resilience in realistic operation scenarios.

Further improvements of smart grid performances may be obtained by predicting the occurrence of critical events, which can affect the power system security and reliability. In this context, the adoption of computational intelligence techniques represents one of the most promising research directions. Armed with such a vision, in paper [5], a knowledge-based data-mining approach, which employs a fuzzy rule-based classification system characterized by a genetically optimized interpretability-accuracy trade-off, is conceptualized for transparent and accurate prediction of decentral smart grid control stability. This paper explores the hierarchy of influence of particular input attributes upon the system stability, analyzing the effect of possible "overlapping" of some input attributes over the other ones from the system stability perspective. A detailed experimental analysis demonstrates the advantages of the proposed approach compared to other stability prediction methodologies proposed in the literature.

Other interesting contributions in this field have been oriented towards identifying the most effective computing paradigms aimed at supporting the large-scale deployment of data-driven smart grid control functions. To address this complex issue, in paper [6], a novel decentralized control paradigm based on dynamic agents is proposed for online smart grid voltage control. This represents a relevant issue to address since the identification of the most effective approach, which is influenced by the available computing resources, and the required control performance, is still an open problem. Detailed simulation results obtained in a realistic case study are presented and discussed to prove the effectiveness and the robustness of the proposed method.

References

1. Heung-gu Son, Y.K.; Kim, S. Time Series Clustering of Electricity Demand for Industrial Areas on Smart Grid. *Energies* **2020**, *13*, 2377.

2. Jiu G.; Wang, Y.; Xie, D.; Zhang, Y. Wind Farm NWP Data Preprocessing Method Based on t-SNE. *Energies* **2019**, *12*, 3622.

3. Jiejie D.; Teng, Y.; Zhang, Z.; Yu, Z.; Sheng, G.; Jiang, X.; Partial Discharge Data Matching Method for GIS Case-Based Reasoning. *Energies* **2019**, *12*, 3677.

4. Ennio B.; Coletta, G.; De Caro, F.; Vaccaro, A.; Villacci, D. Enabling Methodologies for Predictive Power System Resilience Analysis in the Presence of Extreme Wind Gusts. *Energies* **2020**, *13*, 3501.

5. Gorzałczany, M.B.; Piekoszewski, J.; Rudziński, F. A Modern Data-Mining Approach Based on Genetically Optimized Fuzzy Systems for Interpretable and Accurate Smart-Grid Stability Prediction. *Energies* **2020**, *13*, 2559.

6. Andreotti, A.; Petrillo, A.; Santini, S.; Vaccaro, A.; Villacci, D. A Decentralized Architecture Based on Cooperative Dynamic Agents for Online Voltage Regulation in Smart Grids. *Energies* **2019**, *12*, 1386.

Alfredo Vaccaro
Editor

Article

Enabling Methodologies for Predictive Power System Resilience Analysis in the Presence of Extreme Wind Gusts

Ennio Brugnetti, Guido Coletta, Fabrizio De Caro *, Alfredo Vaccaro and Domenico Villacci

Department of Engineering (DING), University of Sannio, 82100 Benevento, Italy;
ennio.brugnetti@unisannio.it (E.B.); gcoletta@unisannio.it (G.C.); vaccaro@unisannio.it (A.V.);
villacci@unisannio.it (D.V.)
* Correspondence: fdecaro@unisannio.it

Received: 23 May 2020; Accepted: 2 July 2020; Published: 7 July 2020

Abstract: Modern power system operation should comply with strictly reliability and security constraints, which aim at guarantee the correct system operation also in the presence of severe internal and external disturbances. Amongst the possible phenomena perturbing correct system operation, the predictive assessment of the impacts induced by extreme weather events has been considered as one of the most critical issues to address, since they can induce multiple, and large-scale system contingencies. In this context, the development of new computing paradigms for resilience analysis has been recognized as a very promising research direction. To address this issue, this paper proposes two methodologies, which are based on Time Varying Markov Chain and Dynamic Bayesian Network, for assessing the system resilience against extreme wind gusts. The main difference between the proposed methodologies and the traditional solution techniques is the improved capability in modelling the occurrence of multiple component faults and repairing, which cannot be neglected in the presence of extreme events, as experienced worldwide by several Transmission System Operators. Several cases studies and benchmark comparisons are presented and discussed in order to demonstrate the effectiveness of the proposed methods in the task of assessing the power system resilience in realistic operation scenarios.

Keywords: power systems resilience; dynamic Bayesian network; Markov model; Probabilistic Modeling; Smart Grid; Resilience Models

---

1. Introduction

The modern electric grids operation policies are based on rigorous reliability and recovery principles, which have been defined in order to allow power systems to operate safely against multiple severe contingencies, providing high quality of electricity supply [1]. In the last decades, due to climate change and environmental temperature increase, extreme weather events are becoming more and more common even in non-tropical regions [2]. Electric networks are particularly vulnerable to these events, which can induce multiple power equipment damages, especially in overhead lines and substations. In this context, the deployment of traditional reliability and restoration-based methodologies could fail in assessing the impacts of these extreme events on power system operation, due to their inability in effectively modelling low-probable but possible fault scenario [3]. To address this complex problem, new computing paradigms based on resilience analysis have been proposed in the literature for reducing the grid vulnerability against severe disturbances, and improving the corresponding restoration strategies [4–10].

Although there is not an universal definition of system resilience, it can be roughly considered as the ability of a system to anticipate and absorb a High Impact Low Probability (HILP) event and regain its normal operating status as quickly as possible [11]. More specifically, according to the UK Energy Research Center [12] the resilience of an electric power system is: "the capacity of an energy system to tolerate disturbance and to continue to deliver affordable energy services to consumers. A resilient energy system can speedily recover from shocks and can provide alternative means of satisfying energy service needs in the event of changed external circumstances".

This definition outlines the need for defining proper indexes for quantifying the system resilience, in order to assess the effectiveness of mitigation strategies reacting to multiple disruptive events. These strategies can be deployed at both planning and operation stage by (i) improving the infrastructural capacity of the power components to withstand extreme stresses; and (ii) reducing the restoration times, by preemptively identifying proper control actions aimed at mitigating the effects of multiple contingencies, and reducing the restoration times. To this aim, the deployment of the traditional N-1 reliability principle does not allow to obtain a reliable analysis in the presence of severe contingencies induced by multiple HILP events. Anyway, evolving from the N-1 to N-k criterion is not a trivial issue to address due to the prohibitive computational costs of considering a wider, and more severe, set of multiple and correlated contingencies. Hence, the employment of probabilistic risk-based approach, which are characterized by relaxed constraints, may be a good trade-off point.

In this context, the development of risk-based methodologies for power system resilience assessment represents a relevant issue to address in order to estimate the actual system vulnerability against HILP events, the expected impacts on system operation, and the effectiveness of the potential countermeasures.

The possible strategies that can be deployed for solving this issue can be classified into two main groups: ex-post and ex-ante analyses. The first class of methods try to infer from operation data related to past outages, the system resilience against each perturbation events over large operation periods. Besides, these methods can contribute to qualitatively identify in which domains the system operator can intervene for increasing the system resilience, e.g., component design, system restoration, network planning and operation.

A different, and more interesting, prospective is offered by ex-ante methods, which aim at identifying preemptive actions, satisfying fixed system resilience requirements. This is a strategic feature, since power systems operators are compelled to reliably predicting the occurrence and the impacts of "extreme events", in order to be able to manage and mitigate their effects on system operation. Moreover, ex-ante methods allow effectively modelling the impacts of various source of uncertainties on system resilience analysis, as far as load forecasts errors, renewable power generators randomness and uncertain power transactions are concerned.

Nowadays, most of the ex-ante methodologies for resilience analysis proposed in the literature are based on Monte Carlo simulations (MCS), which aim at generating synthetic time series representing the system behavior under different weather conditions [4–7,10], and probabilistic techniques based on the minimal path algorithm, which is applied to identify optimal restoration paths based on the definition of "resilience factors" associated at each component [13–17]. Although these methods allow obtaining valuable information about the potential impacts of severe perturbations on system operation, they may fail to model the complex correlations between multiple disruptive events and components fault rates. These limitations mainly derive by the simplified assumptions that need to be assumed in order to make the problem tractable.

To address this complex problem, the deployment of Dynamic Bayesian Networks (DBNs) represents a very promising research direction. These methods allow predicting the impacts of multiple HILP and cascade events on both system operation and restoration, by considering a plurality of possible events and consequences [8,9]. However, the deployment of these DBNs in power system resilience analysis is still at its infancy, and requires further research efforts aimed at developing computational methods for deep simulations, which should be able to assess and compare the correlations of the physical parameters affected by the HILP events with the components fault models, and the corresponding propagation scenarios, from the starting event up to black out and recovery.

Moreover, the computational burden of these methods could be a limiting factor for on-line predictive resilience analysis, which requires problem solutions in very short time-frames, especially if these solutions should be used as input for further computational processes, such as loss of load estimation, and on-line power system contingency analysis.

Finally, new methods aimed at lowering the information granularity of the components fault models in function of their spatial location, and the expected magnitude of the perturbation events are necessary in order to improve the effectiveness of the resiliency analysis, especially for power systems distributed along large geographical area.

On the basis of this literature analysis, it can be argue that the research for new methods aimed at solving the accuracy versus complexity dichotomy in power system resiliency assessment represents a relevant issue to address.

In trying and solving this issue, this paper analyzes the potential role of adaptive probabilistic models for predictive resiliency analysis in the presence of extreme wind gusts, which have been recognized as one of the most critical weather phenomena affecting many European power systems. The adoption of these models allows adapting the component fault parameters in function of the forecast spatial/temporal wind speed evolution, as far as to dynamically estimate the impacts of multiple faults on power system operation, the corresponding worst-case scenario and its occurrence probability. The main innovations of these methods compared to other traditional techniques can be summarized as follows:

1. Differently from the regional approach proposed in [4], a more detailed characterization of wind spatial profiles, which have been acquired by pervasive sensor devices deployed over the lines, has been performed in order to assess their impact on the system components. This leads to resiliency analysis characterized by higher spatial resolution. In particular, the spatial resolution increment in large network disruption analysis is crucial to face adequately with HILP events as described by [18].
2. The analyzed power system is modelled without assuming any simplified network equivalent;
3. Differently from the DBN-based approach proposed in [9], the parameters of the components fault model are correlated to the weather effects;
4. Differently from the approach proposed in [8], the cascade effects induced by multiple components failure are modelled by dynamically adapting the power system topology, which allows lowering the complexity of the assessment procedure.

Several cases studies and benchmark comparisons are presented and discussed in order to demonstrate the effectiveness of the proposed methods in the task of assessing the power system resilience in realistic operation scenario. For each considered case study, a comprehensive scalability analysis is performed in order to assess the computational burden of the proposed methods in function of the system complexity.

2. Mathematical Preliminaries

Predicting the impacts of disruptive events on system operation is a strategic tool for improving the power systems security, since it allows identifying preemptive actions aimed at mitigating the effects of multiple contingencies induced by these severe events [19]. This computing process, which is usually referred as predictive resilience analysis, requires the solution of a set of probabilistic models aimed at (i) characterizing how the disruptive events affect the components failure parameters, (ii) predicting the corresponding components failures, and (iii) assessing their impacts on power system operation. To this aim, modelling techniques based on Markov Process and Bayesian Network represent the most promising enabling methodologies.

2.1. Markov Chain

A Markov Chain is a memory-less discrete stochastic process, satisfying the so called "*Markov property*":

$$P[X_{(t+1)} = x_{(t+1)} | X_{(t)} = x_{(t)}, \dots, X_{(0)} = x_{(0)}] = P[X_{(t+1)} = x_{(t+1)} | X_{(t)} = x_{(t)}] \tag{1}$$

where X is a generic discrete random variable, which assumes a finite number of d possible occurrences (called "*state*") $S_X : \{x_1, \dots, x_d\}$. Equation (1) states that the evolution of the system depends only on the present state and not on the past.

Furthermore, if the following equation holds on the process is called "*homogeneous*":

$$P[X_{(t+1)} = x_j | X_{(t)} = x_i] = P[X_{(h+1)} = x_j | X_{(h)} = x_i] \qquad \forall t, h \in [1, T] \qquad \forall i, j \in [1, d] \tag{2}$$

The latter assures the process being time-invariant, which means that the transition probability matrix $\mathbf{Q}$ has constant parameters over the time. The transition probability matrix is a square matrix of order d:

$$\mathbf{Q} = \begin{bmatrix} q_{11} & q_{1j} & \cdots & q_{1d} \\ q_{i1} & q_{ij} & \cdots & q_{id} \\ \vdots & \vdots & \ddots & \vdots \\ q_{d1} & q_{dj} & \cdots & q_{dd} \end{bmatrix} \tag{3}$$

whose elements q_{ij} represent the conditional probabilities to be in the state j at the time instant t+1 starting from the state i. One of the main properties of this matrix is that the elements of each row have to guarantee that their sum is equal to 1. Hence, once known the state probability vector $\mathbf{x}$ at time instant t, the corresponding probabilities at the next step can be computed as it follows:

$$\mathbf{x}_{(t+1)} = \mathbf{x}_{(t)} \mathbf{Q} \tag{4}$$

The state probability vector at the initial time step is a vector with only one element equal to 1. In case the parameters of $\mathbf{Q}$ change over the time the Markov Chain is called "*time-variant*" and the transition matrix has defined as $\mathbf{Q}(t)$.

2.2. Bayesian Networks

Bayesian Network (BN) is Directed Acyclic Graph (DAG) that allows representing all the casual relationship among a set of correlated variables. The structure of a Bayesian network is based on two main components:

Description 1. *Nodes: represent a set of variables* $S_X : \{X_1, \dots, X_d\}$. *Every node has a conditional probability distribution represented through Conditional Probability Table (CPT).* **Arcs:** *represent the probabilistic dependencies between the variables. A node is called "parent" of a "child" if there is a direct arc connecting the first to the second.*

Each node is characterized by a conditional probability distribution modeled through the Conditional Probability Table (CPT). For each variable X_i, with n parent nodes, $(Pa_1, Pa_2, \dots, Pa_n)$, the CPT is indicated as $P(X_i | Pa_1, Pa_2, \dots, Pa_n)$ and contains all the probability associated to any possible combination between the states of X_i and all its parents.

By using the chain rule, the joint probability distribution of all the BN nodes can be computed as follows:

$$P(S_X) = P(X_1, X_2, \dots, X_d) = \prod_{i=1}^{d} P(X_i | Pa(X_i)) \tag{5}$$

3. Proposed Methodology

The aim of this paper is to propose a computationally effective method for assessing the impacts of severe wind gusts on power system operation by deep simulations of probabilistic models. The final goal is to assess the system reaction to the loss of multiple critical network components, whose failure model parameters are adapted in function of predicted time/spatial wind speed profiles. The latter greatly affect the components fault rate, especially in the presence of extremely high wind speeds, which may damage the overhead line conductors, causing multiple and severe faults, as recently experienced in several European power systems [20]. These severe weather phenomena could interest large geographical area, threatening the correct operation of a large number of power components. Consequently, the number of fault scenarios that should be analyzed may exponentially increase, causing an explosion of the problem cardinality, which needs to be properly managed.

To this aim, two different solution methodologies, which are based on time-varying Markov Chains and Dynamic Bayesian Networks, have been developed and compared.

3.1. Improved Time Varying Markov Chains

As introduced in Section 2.1, a MC is entirely defined by its transition matrix, which owns the information about the probability to evolve from a state to another over the time. If the transition rates are time-dependent, the MC is called time-varying, and the transition probability matrix is variable. Thus, the Equation (4) becomes:

$$\mathbf{x}_{(t+1)} = \mathbf{x}_{(t)}\mathbf{Q}(t) \quad \forall t \in [1, T] \tag{6}$$

where $\mathbf{x}_t$ is the probability state vector at t^{th} time step, whose dimensions are $[1, S]$ with S number of network states, $\mathbf{Q}(t)$ is the time varying transition probability square matrix of order S, whose parameters are time dependents.

This mathematical tool could play an important role in predictive resilience analysis, since it allows describing the impacts of the time/spatial wind speed profiles on the fault and reparation probability of each network component. To this aim, each time-varying MC state represents a possible power system operation state, hence obtaining a number of $S = b^n$ possible states, where n is the number of critical power components, and b is the number of their operation states. Without loss of generality, two operation states are considered for describing power component operation, namely, "*Run*" (the component is in service) or "*Fault*" (the component is out of service). Hence, a generic power system operation state s_i, at each time step, is described by an unique combination of components operating conditions, as shown in the following example for a network with two critical components ($n = 2$):

$$s_i : \{L_{1R}L_{2R} ; L_{1F}L_{2R} ; L_{1R}L_{2F} ; L_{1F}L_{2F}\} \tag{7}$$

The elements of the transition probability matrix in a discrete Markov Chain depends on the probability rates to evolve from any state s_i to all the others, where the sum of all transition probability rates has to be unitary for each row of $\mathbf{Q}$. In particular, the starting and arrival states are organized by rows and columns of $\mathbf{Q}$, which assumes the following generalization form:

$$\mathbf{Q}(t) = \begin{bmatrix} q_{11}(t) & q_{1j}(t) & \cdots & q_{1S}(t) \\ q_{i1}(t) & q_{ij}(t) & \cdots & q_{iS}(t) \\ \vdots & \vdots & \ddots & \vdots \\ q_{S1}(t) & q_{Sj}(t) & \cdots & q_{SS}(t) \end{bmatrix} \tag{8}$$

where $q_{ij}(t)$ can be computed as:

$$q_{ij}(t) = \prod_{k=1}^{n} c_t^{(k)} \quad \forall t \in [1, T] \tag{9}$$

where $c_t^{(k)}$, which depends on the characteristic of the k-th component state transition, can be computed as follows:

$$
c_t^{(k)} = \begin{cases}
\lambda_t^{(k)} & \text{if the k-th component moves from } \textit{Run} \text{ to } \textit{Fault} \text{ state} \\
\mu_t^{(k)} = \left(1 - \lambda_t^{(k)}\right) & \text{if the k-th component remains in } \textit{Run} \text{ state} \\
\gamma_t^{(k)} & \text{if the k-th component moves from } \textit{Fault} \text{ to } \textit{Run} \text{ state} \\
\phi_t^{(k)} = \left(1 - \gamma_t^{(k)}\right) & \text{if the k-th component remains in } \textit{Fault} \text{ state}
\end{cases}
\tag{10}
$$

where $\lambda_t^{(k)}$ and $\gamma_t^{(k)}$ are the time-varying fault and restoration rates of the k-th component, respectively. The variation of these parameters in time reflects the influence of the time/spatial wind speed evolution on the fault and reparation probability of the k-th component. In this context, the following piece-wise approximation of the component fragility curve is assumed as the main driving factor affecting the fault rate [21]:

$$
\lambda_t^{(k)} = \begin{cases}
0 & w_t^{(k)} < w_{(1,k)} \\
f\left(w_t^{(k)}\right) & w_{(1,k)} \leq w_t^{(k)} < w_{(2,k)} \\
1 & w_t^{(k)} \geq w_{(2,k)}
\end{cases}
\tag{11}
$$

where $w_t^{(k)}$ is the wind speed expected on the k-th component, while $w_{(1,k)}$ and $w_{(2,k)}$ are static thresholds, which should be properly identified in order to approximate the component fragility curve by a piece-wise linear function.

As far as the component reparation rate is concerned, it can be computed based on the mean time to repair the k-th component, as follows:

$$
\gamma^{(k)} = 1/\tau^{(k)}
\tag{12}
$$

This simplified assumption is based on the fact that, on the average, every τ^k time steps the k-th component is expected to be repaired, hence its restoration probability can be reasonably expressed as shown in Equation (12). Roughly speaking, this assumption considers a uniform probability distribution for the event *the k-th component is repaired* over the time. However, it is important to outline that also the mean time to repair is correlated to the weather condition insisting on the power component. To this aim, more advanced techniques should be used for modeling this behaviour by defining a "repair curve" defined similarly to the "fragility curve". This problem is currently under investigation by the Authors.

Once the time-variant transition probability matrix has been updated based on the expected evolution of the wind speed time/spatial profiles, the state probability for a step ahead $(t + 1)$ can be computed by using Equation (9).

3.2. Dynamic Bayesian Network

A different, and more effective, methodology for predictive resilience analysis is based on the development of a Dynamic Bayesian Network, which allows modelling time-varying systems based on cause-effect relationships modeled through DAG. The main difference with respect to traditional BN is that each node at time step t can be affected by the state of the Parents (Pa) variables through inter-slice connections.

The construction of CPT matrix for each couple of Child–Parents relationship is the core of the proposed DBN model, which describes the operation state of each power component.

The flow scheme of the DBN is reported in Figure 1, showing the operation state of the k-th component at time step t, which depends on both the previous operations state at $(t - 1)$ and the expected wind speed at time step t. In particular, similarly to the MC paradigm, the binary random variable "*state*" could assume the states *Run* and *Fault*, and the random variable "*wind*" varies in proper intervals, depending by the expected wind speed on the k-th component at time step t. This feature allows modeling the impacts on the power components induced by the expected time/spatial wind speed profiles. Hence, the variable *"wind"* assumes three possible occurrences as shown in Table 1, where the values $a_t^{(\{1,2,3\},k)}$ are the occurrence probabilities characterizing a generic class interval. The wind speed discretization, which has been performed by applying Equation (11), is necessary in order to model this process in the DBN context.

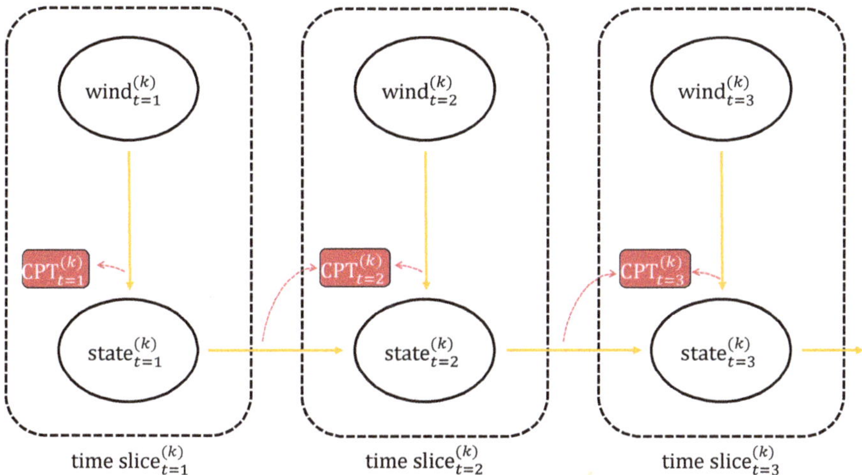

Figure 1. Flow scheme of the proposed DBN.

Table 1. Probability of wind speed.

$P_t^{(k)}$	Weak $w_t^{(k)} < w_{(1,k)}$	Medium $w_{(1,k)} \leq w_t^{(k)} < w_{(2,k)}$	Strong $w_t^{(k)} \geq w_{(2,k)}$
wind	$a_t^{(1,k)}$	$a_t^{(2,k)}$	$a_t^{(3,k)}$

In particular, the integration of this random variable in the proposed DBN has been obtained by clustering the piecewise fragility curve in three parts, as shown in Figure 2b. Thus, the classification *"weak"*, *"medium"*, and *"strong"* indicates that the impact of the wind gust on the k-th component.

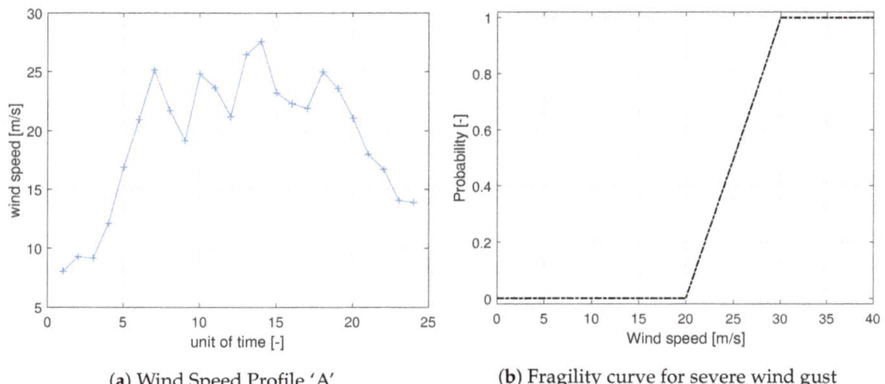

(a) Wind Speed Profile 'A'   (b) Fragility curve for severe wind gust

Figure 2. Failure modelling.

The proposed DBN model has been developed by defining two CPTs, one modelling the relation between the wind and the single component state, and the other describing the joint effect of the previous component state, and the current value of the wind speed, on the current component state. These CPTs can be expressed as follows:

Tables 2 and 3 describe the DBN conditional probabilities for the k-th critical component, whose parameters are computed according to Equations (11) and (12). Then, by using the total probability law, the component state probabilities at time k are computed as indicated in Equation (13). In particular, the latter equation is specific for the case ($\textbf{state}_{(t)} = Run$), but the replacement of the latter term with ($\textbf{state}_{(t)} = Fault$) in the same equation allows computing the corresponding fault state probability.

Table 2. CPT for $t = 1$.

$\textbf{CPT}^{(k)}_{(t=1)}$		State	
		Run	**Fault**
wind	weak	$\mu_t^{(k)}(w_t^{(k)})$	$\lambda_t^{(k)}(w_t^{(k)})$
	medium	$\mu_t^{(k)}(w_t^{(k)})$	$\lambda_t^{(k)}(w_t^{(k)})$
	strong	$\mu_t^{(k)}(w_t^{(k)})$	$\lambda_t^{(k)}(w_t^{(k)})$

Table 3. CPT for $t > 1$.

$\textbf{CPT}^{(k)}_{(t>1)}$			$\textbf{State}_{(t)}$	
			Run	**Fault**
$\textbf{state}_{(t-1)} \cup \textbf{wind}_{(t)}$	Run	weak	$\mu_t^{(k)}(w_t^{(k)})$	$\lambda_t^{(k)}(w_t^{(k)})$
	Run	medium	$\mu_t^{(k)}(w_t^{(k)})$	$\lambda_t^{(k)}(w_t^{(k)})$
	Run	strong	$\mu_t^{(k)}(w_t^{(k)})$	$\lambda_t^{(k)}(w_t^{(k)})$
	Fault	weak	$\gamma_t^{(k)}$	$\phi_t^{(k)}$
	Fault	medium	$\gamma_t^{(k)}$	$\phi_t^{(k)}$
	Fault	strong	$\gamma_t^{(k)}$	$\phi_t^{(k)}$

It is worth noting that Equation (13) takes into account the occurrence probabilities for each wind speed class interval, which is a very useful information, considering that the expected time evolution of the wind speed at the k-th component can be preliminary inferred from pre-processing data analysis, and it can be considered as an input variable for the DBN.

However, for each t, the wind speed probability vector will be a null vector, having only a single element equal to 1. The latter is the corresponding class interval characterizing the predicted wind speed value at time step t. For this reason Equation (13) is simplified, since only the expected wind speed probability class is not null. This leads to Equations (14) and (15), in which the term w_x indicates the occurred wind speed class at time t.

The probabilities describing the components state operation over the time, namely *Run* and *Fault*, are collected in the tensor $\textbf{Y}[T, n, b]$, which allows computing the network state probabilities through a multiplication over the n critical components by considering all their operation states. To this aim, the hypothesis of statistical independence between the faults/restorations of the critical components has been assumed [22].

This computing process, which has described by Algorithm 1, requires a proper indexing of $\textbf{Y}[T, n, b]$, which considers all the possible combinations of the components operation states. The following explanatory example, which considers the case of two critical components described in Equation (7), could be useful to clarify this concept:

$$
\begin{aligned}
P\left[Run^{(k)}_{(t)}\right] =\ & P\left[Run^{(k)}_{(t)}\mid Run^{(k)}_{(t-1)}\cup weak^{(k)}_{(t)}\right]\cdot P\left[Run^{(k)}_{(t-1)}\right]\cdot P\left[weak^{(k)}_{(t)}\right]+\\
& +P\left[Run^{(k)}_{(t)}\mid Run^{(k)}_{(t-1)}\cup medium^{(k)}_{(t)}\right]\cdot P\left[Run^{(k)}_{(t-1)}\right]\cdot P\left[medium^{(k)}_{(t)}\right]+\\
& +P\left[Run^{(k)}_{(t)}\mid Run^{(k)}_{(t-1)}\cup strong^{(k)}_{(t)}\right]\cdot P\left[Run^{(k)}_{(t-1)}\right]\cdot P\left[strong^{(k)}_{(t)}\right]+\\
& +P\left[Run^{(k)}_{(t)}\mid Fault^{(k)}_{(t-1)}\cup weak^{(k)}_{(t)}\right]\cdot P\left[Fault^{(k)}_{(t-1)}\right]\cdot P\left[weak^{(k)}_{(t)}\right]+\\
& +P\left[Run^{(k)}_{(t)}\mid Fault^{(k)}_{(t-1)}\cup medium^{(k)}_{(t)}\right]\cdot P\left[Fault^{(k)}_{(t-1)}\right]\cdot P\left[medium^{(k)}_{(t)}\right]+\\
& +P\left[Run^{(k)}_{(t)}\mid Fault^{(k)}_{(t-1)}\cup strong^{(k)}_{(t)}\right]\cdot P\left[Fault^{(k)}_{(t-1)}\right]\cdot P\left[strong^{(k)}_{(t)}\right]
\end{aligned}
\tag{13}
$$

$$
\begin{aligned}
P\left[Run^{(k)}_{(t)}\right] =\ & P\left[Run^{(k)}_{(t)}\mid Run^{(k)}_{(t-1)}\cup w_{x(t)}^{(k)}\right]\cdot P\left[Run^{(k)}_{(t-1)}\right]\cdot P\left[w_{x(t)}^{(k)}\right]+\\
& +P\left[Run^{(k)}_{(t)}\mid Fault^{(k)}_{(t-1)}\cup w_{x(t)}^{(k)}\right]\cdot P\left[Fault^{(k)}_{(t-1)}\right]\cdot P\left[w_{x(t)}^{(k)}\right]=\\
=\ & \mu_t^k\cdot P\left[Run^{(k)}_{(t-1)}\right]+\gamma_t^k\cdot P\left[Fault^{(k)}_{(t-1)}\right]
\end{aligned}
\tag{14}
$$

$$
\begin{aligned}
P\left[Fault^{(k)}_{(t)}\right] =\ & P\left[Fault^{(k)}_{(t)}\mid Run^{(k)}_{(t-1)}\cup w_{x(t)}^{(k)}\right]\cdot P\left[Run^{(k)}_{(t-1)}\right]\cdot P\left[w_{x(t)}^{(k)}\right]+\\
& +P\left[Fault^{(k)}_{(t-1)}\mid Fault^{(k)}_{(t-1)}\cup w_{x(t)}^{(k)}\right]\cdot P\left[Fault^{(k)}_{(t-1)}\right]\cdot P\left[w_{x(t)}^{(k)}\right]=\\
=\ & \lambda_t^k\cdot P\left[Run^{(k)}_{(t-1)}\right]+\phi_t^k\cdot P\left[Fault^{(k)}_{(t-1)}\right]
\end{aligned}
\tag{15}
$$

$$
\begin{cases}
P[L_{1R}\,L_{2R}]_{(t)} & \rightarrow \mathbf{y}[1,t]=\mathbf{Y}[t,1,1]\cdot\mathbf{Y}[t,2,1]\\
P[L_{1R}\,L_{2F}]_{(t)} & \rightarrow \mathbf{y}[2,t]=\mathbf{Y}[t,1,1]\cdot\mathbf{Y}[t,2,2]\\
P[L_{1F}\,L_{2R}]_{(t)} & \rightarrow \mathbf{y}[3,t]=\mathbf{Y}[t,1,2]\cdot\mathbf{Y}[t,2,1]\\
P[L_{1F}\,L_{2F}]_{(t)} & \rightarrow \mathbf{y}[4,t]=\mathbf{Y}[t,1,2]\cdot\mathbf{Y}[t,2,2]
\end{cases}
\tag{16}
$$

Algorithm 1 Network state probabilities computing.

```
 1: load the label tag for all S network operation states (a.e. < 10 21 > , < 11, 20 >, … )
 2: loop
 3:    move over each network operation state (s_i)
 4:    loop
 5:       move over the time (t)
 6:       loop
 7:          move over each network line
 8:          load the k-th line tag (1, …, n) from s_i
 9:          load the occurred operative state tag (run/fault) for each k-th line from s_i
10:          get from tensor Y the probability for the extracted 'line-state-time step' tag set
11:          store the probability in the t-th cell of a temporary array and scroll through it
12:       end loop
13:       compute the product of all temporary array elements
14:       store the i-th network state probability at time t (s_i) in matrix y[i, t]
15:       erase the temporary array
16:    end loop
17: end loop
```

3.3. Effect of the Time-Step Choice in the Developed Discrete-Time Models

Since both proposed methodologies are discrete-time models it is important to analyze the effect of time-step choice on the modeled system. In particular, the time step does not affect the dynamics of both Bayesian Network and Markov Time Varying Process but it affects the input data of the proposed methodologies such as:

- line maintenance probabilities after a fault: they need to be adapted by considering the adopted time step;
- wind speeds: where the predicted values are usually the mean value for that time interval. Thus, a shorter time interval implies greater volatility as well as an excessively large time interval cannot take into account the wind fluctuations. The right choice might be based on the base of wind frequency of the severe events of the area under analysis.

4. Case Studies

The proposed methodologies have been applied in order to perform predictive resilience analyses for several networks, which are characterized by different lines number. The final goal is to compare the performances of the proposed non-homogeneous Improved Markov chains and the Dynamic Bayesian network-based approaches, in terms of both accuracy and computation burden.

In particular, since the weather conditions can sensibly vary along the transmission line route, hence affecting the corresponding fault model parameters to a considerable extent. Consequently, worst-case assumption has been adopted by considering, for each time step, the maximum wind speed predicted along the line route as input to the fragility curve.

According to this approach, the failure rate of the critical lines have been modeled through the "fragility curve" reported in [21], which, at each time step, allows computing the line failure rates in function of the maximum wind speed expected at the lines locations.

4.1. Numerical Results

The proposed methodologies have been tested on several case studies characterized by an increasing number of critical lines and different spatial wind profiles. For the sake of simplicity, several simplified hypothesis have been assumed during these studies. First of all, the wind speed in each conductor's surrounding is the same, with a profile as shown in Figure 2a. This hypothesis could compromise the results effectiveness, since severe extreme wind gusts usually propagates as fronts, especially in wide-area power systems. However, the methodologies proposed in this paper are designed to model the impacts of spatial and temporal wind speed profiles, which can be estimated by high-resolution wind speed forecasting, as far as to consider the impacts of these profiles on the failure and recovery rates associated to each system component. Moreover, the component failure rate is modeled based on the fragility curve depicted in Figure 2b, which can be used to compute the time-varying transition matrices of the TVMC model, and the CPTs describing the DBN, as described in Sections 3.1 and 3.2.

4.2. Predictive Resilience Analysis: TVMC vs. DBN Approach

This section shows the comparison of TVMC and DBN approaches for power system predictive analysis. In particular, Figures 3–5 show the probability profiles obtained by applying the Markov Process (left side) and the DBN (right side) methodology for one, two and three critical lines, respectively. In particular, each profile represents the probability of the network to be in a certain state, which is characterized by the code shown in the legend (e.g., <10 21 31 40 51>). The latter codifies the information about the state of each line in the system: the first identifies the line, while the second indicates its state (1 the line is "in service", 0 the line is "out of service"). For example, considering a network with five critical components, the code <10 21 31 40 51> identifies the network state in which lines 1 and 4 are "out of service" and lines 2, 3 and 5 are "in service". This representation is more convenient than the semantic one used in the previous sections, since it allows automatically generating the states enumeration.

It is worth noting that in Figures 3 and 5 there are states with identical probability of occurrence. This behaviour can be reasonably justified by recalling that the proposed case studies consider the same weather conditions on all the critical lines in the network; hence, the network states characterized by the same ratio between "in service" and "out of service" lines cannot be distinguished each other.

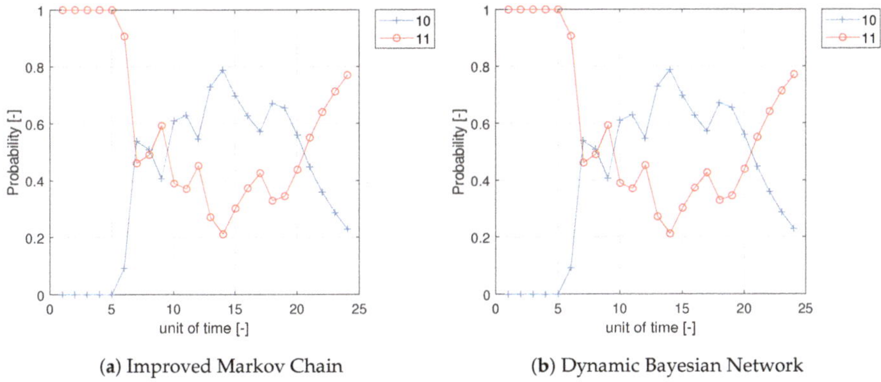

(**a**) Improved Markov Chain   (**b**) Dynamic Bayesian Network

Figure 3. Network State Probability: One critical line with wind speed profile 'A'.

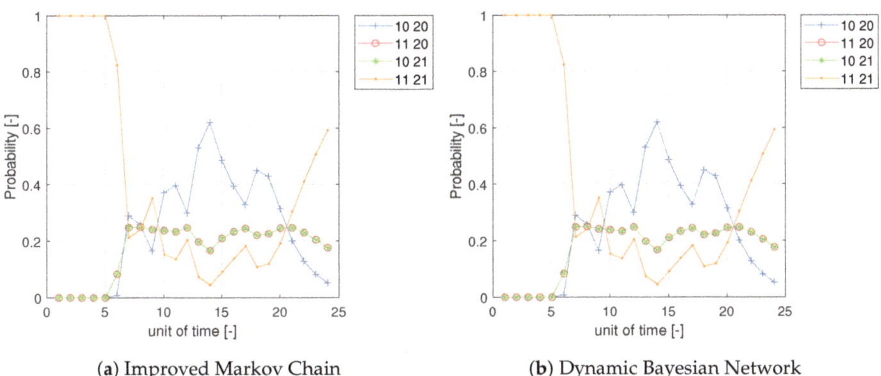

(**a**) Improved Markov Chain   (**b**) Dynamic Bayesian Network

Figure 4. Network State Probability: Two critical lines with wind speed profile 'A'.

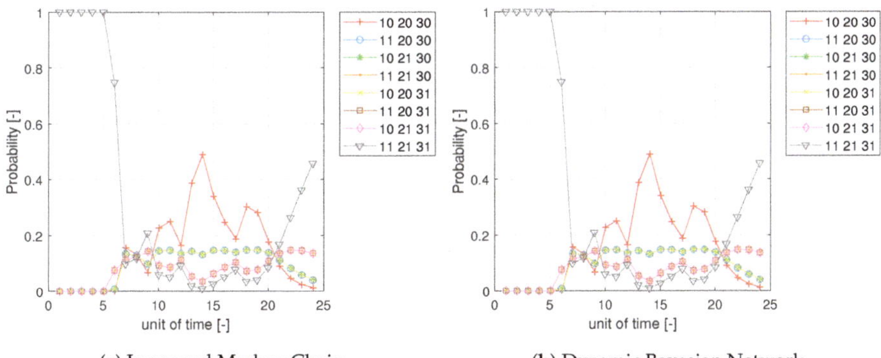

(**a**) Improved Markov Chain   (**b**) Dynamic Bayesian Network

Figure 5. *Cont.*

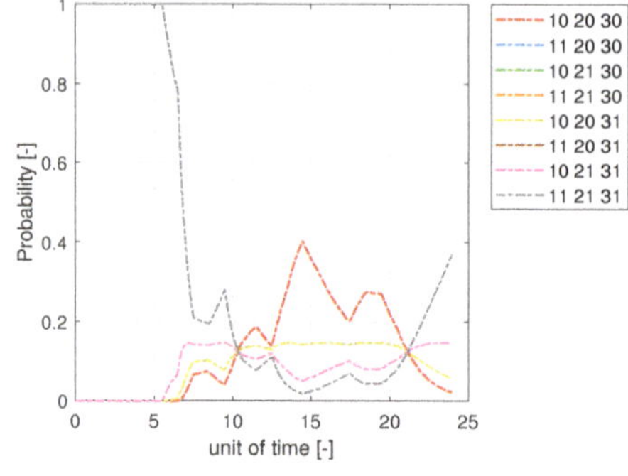

(c) Standard Markov Model

Figure 5. Network State Probability: Three critical lines with wind speed profile 'A'.

The results of the proposed methodologies are proven to be equal as depicted in Figures 3 and 5. This result is not unexpected because DBNs are a particular instance of Markov Process, and, under certain conditions (i.e., global and local Markov Properties), they allow obtaining the same results [23]. In this context, given the structure of the proposed DBN, it is proved that this equivalence is assured by only satisfying of the local direct Markov property, which is a prerequisite to apply the chain rule (5).

Furthermore, in order to outline the main benefits of the proposed methodologies a standard reliability model, which are based on a time-continuous Markov chain for a 3 line-grid as shown in Figure 5, has been compared.

The main difference between the proposed methodologies and this modeling technique, which is widely adopted for power system reliability analysis, is that the neglects of the occurrence of multiple component faults and repairing. This assumption is well verified in reliability analysis, but it could be not suitable for resilience analysis, since the probability of multiple failures in the presence of extreme events is not negligible, as experienced worldwide by several Transmission System Operators. Hence, this assumption may lead to underestimate the probability of the worst operation states, in the presence of high speed gusts. Hence, the reliability standard model assumptions may lead to underestimate the probability of the worst operation states (<10 20 30>) in the presence of high speed gusts as shown in Figure 5c.

Moreover, for the sake of completeness, both methodologies have been tested with a further wind spatial profile ('B'), which is characterized by greater values magnitude than the previous case ('A'), where some occurrences lie on the 100% "out of service" probability side of the fragility curve.

In particular, a 2 line-network has been considered in this case, where the wind speed profiles are depicted in Figure 6a, where, for $t > 14$, the wind speed is greater than the maximum strength limit going to cause the likely failure of the overhead line (Figure 2b). Indeed, the full operation grid state (<11 21>) dramatically drops to 0 due to the occurred severe wind speeds as shown by Figure 7a,b.

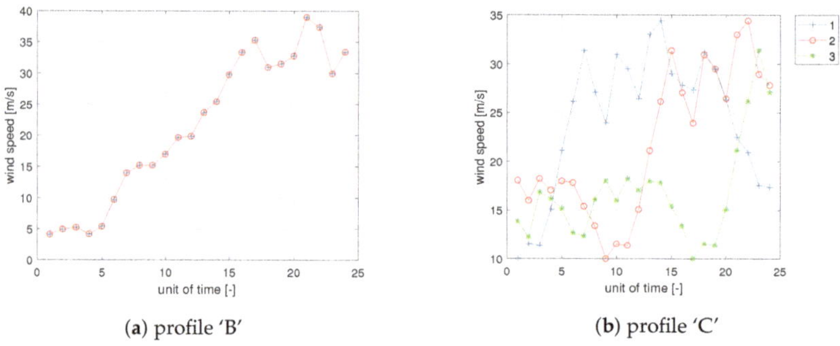

(a) profile 'B'　　　　　　　　　　　　(b) profile 'C'

Figure 6. Wind speed spatial profiles.

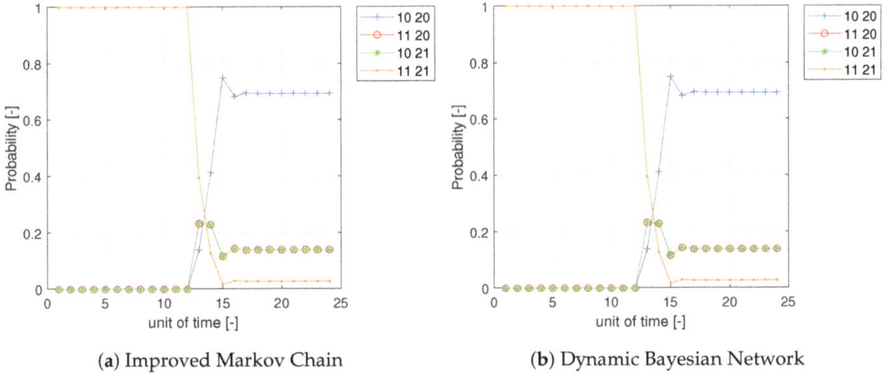

(**a**) Improved Markov Chain   (**b**) Dynamic Bayesian Network

Figure 7. Network State Probability: Two critical lines with the wind spatial profile 'B'.

4.3. Spatial Wind Profile Characterization

The development of the proposed methodologies is based on the spatial characterization of the wind profile at high resolution, where the latter may permit to asses the impact of a weather perturbation evolution over the grid. Therefore, a further case study has been developed, where the latter is based on the spatial wind profile ('C') (Figure 6b) and a 3-line grid, in order to support this statement.

In particular, for the sake of clarity, the spatial wind profiles shown in Figure 6b have been generated from the same wind profile lagged over the time of 8 units. This has been realized for better highlighting the effect of the spatial effect on the network probability states.

Indeed, Figures 6b and 8 reveal how the weather perturbation movement dramatically affects the states probability profiles evolution. In particular, the combined analysis of the latter figures has revealed that for:

- $t < 5$: the spatial average wind speed is low over the grid, therefore the most likely state is that concerning a full operative condition (<11 21 31>);
- $5 \leq t < 14$: line 1 is affected by a severe weather event with increasing wind gust speeds. The most likely state becomes that with line 1 out of service (<10 21 31>);
- $14 \leq t < 22$: The severe weather event moves from line 1 to 2. Since line 1 is likely under maintenance and line 2 is interested by strong wind gusts the most likely state is (<10 20 31>);
- $t \geq 22$: line 1 may be repaired and the storm is affecting line 3 now. Hence the most probable state is (<10 20 30>). Furthermore, the full operative grid state is the least probable because the grid has been not completely repaired yet.

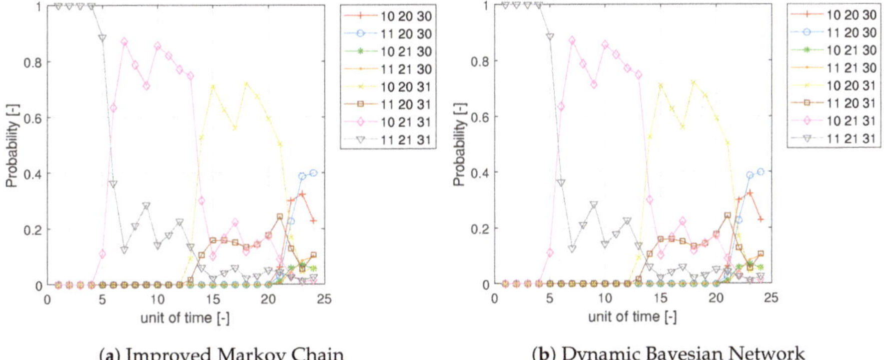

(**a**) Improved Markov Chain   (**b**) Dynamic Bayesian Network

Figure 8. Network State Probability: Two critical lines with the wind spatial profile 'C'.

4.4. Analysis on the Computational Efforts

In consideration of the previous case studies it is worth observing that one of the main problems when analyzing multiple contingency scenarios derives from the high number of possible state to be analyzed. In particular, in the presence of n critical components, the possible combination are 2^n. This exponential growth is the main issue to address in performing predictive resilience analysis for complex power systems. This is confirmed in Table 4, which shows the computation burden required by TVMC and DBN approaches as the critical lines number increases. Clearly it is possible to note that, for both methodologies, the computation burden exponentially increases with the number of possible network states. Even though the TVMC-based approach is more effective for small test networks, DBN methodology allows to effectively solve the problem even when TVMC fails. In fact it can be observed that DBN allows analyzing networks up to 25 critical lines, with respect to 11 critical lines, which is the limit of TVMC approach. Hence, DBN can almost double the capability of the more intuitive TVMC without any loss of information and accuracy.

Table 4. Table Type Styles.

	Improved Markov Process		**DBN**
Critical Lines Number	**Transition Matrix Dimension**	**Computation Time [s]**	**Computation Time [s]**
1 Line	2×2	0.0008	0.1131
2 Lines	4×4	0.001	0.1410
3 Lines	9×9	0.010	0.1448
4 Lines	16×16	0.012	0.1516
5 Lines	32×32	0.017	0.1663
6 Lines	64×64	0.083	0.1766
7 Lines	128×128	0.102	0.2189
8 Lines	256×256	0.400	0.1839
9 Lines	512×512	2.010	0.2016
10 Lines	1024×1024	8.664	0.3408
11 Lines	2048×2048	49.79	0.3918
12 Lines	-	-	0.8164
15 Lines	-	-	3.160
20 Lines	-	-	86.18
25 Lines	-	-	3316

The previous analysis has neglected the impact of the time resolution and forecasting horizon on the computational effort. In fact, the latter is not affected by them because the most part of computational burden is related to state generation and not to the basic arithmetic operators required for computing the model evolution. Hence, for the sake of completeness, a case study on a long run simulation has been performed by applying the DBN model, where the spatial wind profiles and the probability state evolution have been depicted in Figure 9a,b, respectively. In particular both figures allow to observe the full restoration of the system after a certain time, which is necessary to complete all recovery operation after a severe weather event over 2 line-grid (<11 21>).

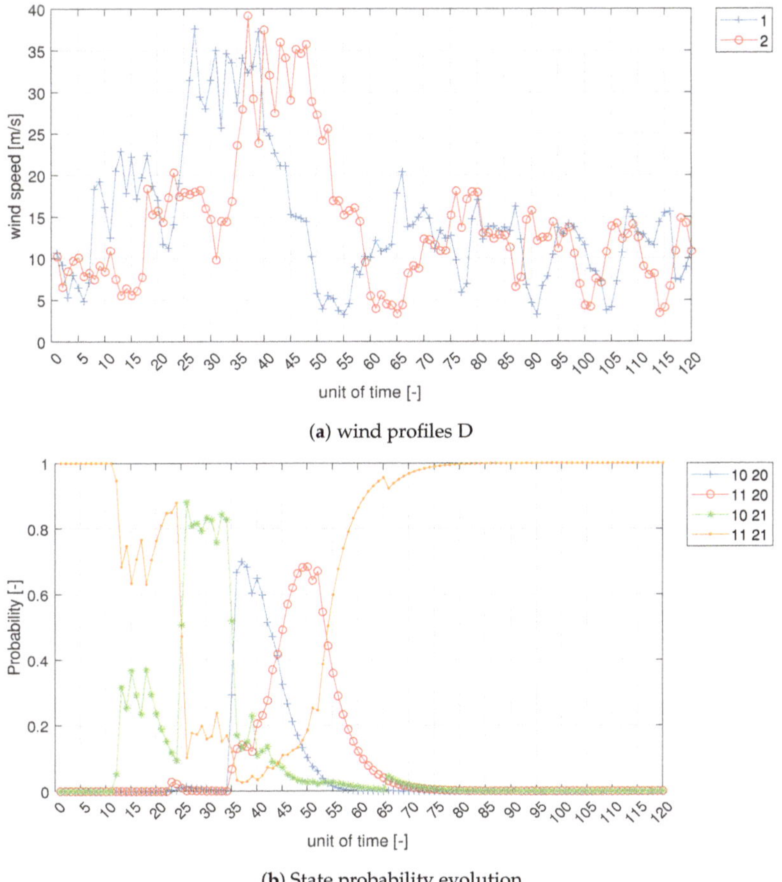

(**a**) wind profiles D

(**b**) State probability evolution

Figure 9. Long Run Simulation DBN-based.

5. Future Research Directions

The proposed methodologies do not take into account the randomness of the time/spatial wind speed profiles, which have been considered as an input of the resilience assessment process. This is a relevant issue to address, since this information is generated by forecasting algorithms, which could be characterized by strong uncertainties, especially in the presence of medium-term forecasting horizons. These uncertainties should be properly considered in solving both the TVMC and DBN models, since they can affect the computed state probabilities to a considerable extent.

In the Authors' opinion, which have been confirmed by some preliminary results, it can be argue that a greater impact occurs when the wind speed forecasting errors lie around the wind class interval value of the fragility curve, which dramatically change the component fault ratios.

A further point of analysis concerns the not spatial uniformity in weather conditions over the transmission lines. A way to consider the spatial weather effect over the network is following an approach similar to as done in wind spatial forecasting where, given a horizontal spatial resolution, each grid cell is characterized by a spatial average wind value. Thus, by overlaying this grid with the network scheme is possible relating a wind value to each line part as shown in Figure 10a.

Obviously, a further improvement may be considering a line as split in several parts as shown by Figure 10b, where the overall operation condition probability of the whole line is given by the product of the operation probabilities of each line part. Hence, in this scenario the greater flexibility of DBN may allow to better manage possible high cardinality issues than the Improved Markov Chain.

Another relevant issue to address deals with the impacts of the weather conditions on the components restoring probabilities, which can greatly slowing down the repairing times, especially in the presence of extreme weather conditions. To address this issue, DBN-based approaches seem to be the most effective solution strategy, since they allow to model the time/spatial wind speed randomness, and the corresponding impacts on both fault and restoration rates in a very effective way.

A further analyzed issue, which is currently under investigation by the Authors, is the improvement of the proposed DBN-based methodology by integrating adaptive fault models for specific network components, as far as tower, transformers, and primary station are concerned. This could be obtained by properly modeling the cause-effect relation between the considered components.

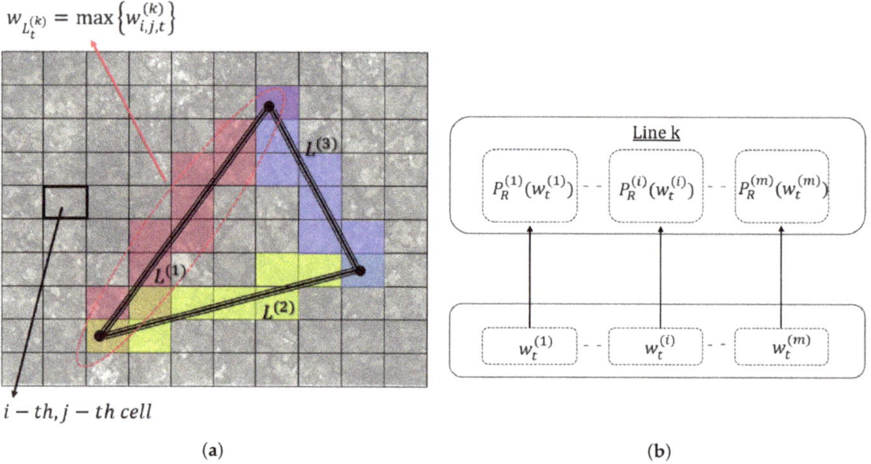

Figure 10. Proposed and future trend-approaches in transmission line modeling. (**a**) Wind spatial characterization on the network. Dotted red circle: proposed approach. Colored cell: Finer mesh approach. (**b**) Piece-wise line model with finer mesh approach.

6. Conclusions

Predictive resilience analysis is assuming a major role in modern power system operation, where the strive for strictly reliability and security constraints is pushing system operators to identify effective strategies aimed at reducing the grid vulnerability against severe disturbances, and improving the corresponding restoration strategies.

In trying and solving this issue, this paper proposed two methodologies, which are based on Time Varying Improved Markov Chain and Dynamic Bayesian Network, for assessing the system resilience against extreme wind gusts. The main idea was to employ lines weather dependent fault parameters in function of the forecast spatial/temporal wind speed evolution, and to dynamically estimate the impacts of multiple faults on power system operation.

Simulation results obtained on several operation scenarios confirmed the effectiveness of the proposed methods in the task of estimating the system resilience by a deep simulation of all the possible network states, which were generated by considering all the possible fault scenario. The proposed methodologies have been mutually compared, and benchmarked with a traditional reliability modelling technique. On the basis of these results, the following conclusions can be drawn: (i) The proposed Markov Chain and Dynamic Bayesian Network-based techniques allow effectively modelling the effects of multiple disruptions and restorations, which are not infrequent in the presence of extreme weather conditions. (ii) Compared to the Markov Chain-based techniques, the Dynamic

Bayesian Network is characterized by lower computational burden. In particular, the obtained results have revealed the effectiveness of DBN in facing with high cardinality problems, which mainly derives from the low complexity of the solution algorithm, that does not require the solution of ordinary differential equations, and complex matrix manipulations. Furthermore, the Dynamic Bayesian Network is characterized by an improved capability in modelling the statistical dependencies characterizing the random processes influencing the fault models and the restoring operations, which is one of the direction of our future research activities.

Author Contributions: A.V. conceived the proposal, E.B. and F.D.C. conceived the network model based on Dynamic Bayesian network. E.B., G.C., and F.D.C. conceived the network model based on Markov Process. E.B. wrote the introductory section, E.B. and F.D.C. the Mathematical Preliminaries, F.D.C. wrote the Proposed Methodology, G.C. and F.D.C. the Case Studies and Results sections. A.V. wrote the rest of the manuscript. E.B., F.D.C., and A.V. improved the manuscript after the reviewer response. All the authors contributed to the interpretation of the results. A.V. and D.V. took the lead in writing the manuscript. All authors provided critical feedback and helped shape the research, analysis and manuscript. All authors have read and agreed to the published version of the manuscript.

Funding: The author(s) received no financial support for the research, authorship, and/or publication of this article.

Conflicts of Interest: The authors declare no conflict of interest.

Abbreviations

The following abbreviations are used in this manuscript:

CPT	Conditional Probability Table
DAG	Direct Acyclic Graph
HILP	High Impact Low Probability
MC	Markov Chain
MCS	Monte Carlo Simulation
BN	Bayesian Network
DBN	Dynamic Bayesian Network
TVMC	improved Time Varying Markov Chain
DAG	Directed Acyclic Graph

References

1. Tabatabaei, N.M.; Ravadanegh, S.N.; Bizon, N. (Eds.) *Power Systems Resilience Modeling, Analysis and Practice*; Springer: Berlin/Heidelberg, Germany, 2019.
2. Miglietta, M.M.; Mazon, J.; Motola, V.; Pasini, A. Effect of a positive Sea Surface Temperature anomaly on a Mediterranean tornadic supercell. *Sci. Rep.* **2017**, *7*, 1–8.
3. Heylen, E.; Ovaere, M.; Proost, S.; Deconinck, G.; Van Hertem, D. A multi-dimensional analysis of reliability criteria: From deterministic N-1 to a probabilistic approach. *Electr. Power Syst. Res.* **2019**, *167*, 290–300. [CrossRef]
4. Panteli, M.; Mancarella, P. Modeling and evaluating the resilience of critical electrical power infrastructure to extreme weather events. *IEEE Syst. J.* **2015**, *11*, 1733–1742. [CrossRef]
5. Espinoza, S.; Panteli, M.; Mancarella, P.; Rudnick, H. Multi-phase assessment and adaptation of power systems resilience to natural hazards. *Electr. Power Syst. Res.* **2016**, *136*, 352–361. [CrossRef]
6. Fu, G.; Wilkinson, S.; Dawson, R.J.; Fowler, H.J.; Kilsby, C.; Panteli, M.; Mancarella, P. Integrated approach to assess the resilience of future electricity infrastructure networks to climate hazards. *IEEE Syst. J.* **2017**, *12*, 3169–3180. [CrossRef]
7. Navarro-Espinosa, A.; Moreno, R.; Lagos, T.; Ordoñez, F.; Sacaan, R.; Espinoza, S.; Rudnick, H. Improving distribution network resilience against earthquakes. In Proceedings of the IET International Conference on Resilience of Transmission and Distribution Networks (RTDN 2017), Birmingham, UK, 26–28 September 2017.
8. Codetta-Raiteri, D.; Bobbio, A.; Montani, S.; Portinale, L. A dynamic Bayesian network based framework to evaluate cascading effects in a power grid. *Eng. Appl. Artif. Intell.* **2012**, *25*, 683–697. [CrossRef]

9. Yodo, N.; Wang, P.; Zhou, Z. Predictive resilience analysis of complex systems using dynamic Bayesian networks. *IEEE Trans. Reliab.* **2017**, *66*, 761–770. [CrossRef]

10. Li, Y.; Xie, K.; Wang, L.; Xiang, Y. Exploiting network topology optimization and demand side management to improve bulk power system resilience under windstorms. *Electr. Power Syst. Res.* **2019**, *171*, 127–140. [CrossRef]

11. Mohamed, M.A.; Chen, T.; Su, W.; Jin, T. Proactive Resilience of Power Systems Against Natural Disasters: A Literature Review. *IEEE Access* **2019**, *7*, 163778–163795. [CrossRef]

12. Chaudry, M.; Ekins, P.; Ramachandran, K.; Shakoor, A.; Skea, J.; Strbac, G.; Wang, X.; Whitaker, J. *Building a Resilient UK Energy System: Working Paper*; UK Energy Research Centre (UKERC): London, UK, 2009.

13. Khalil, Y.; El-Azab, R.; Adma, M.; Elmasry, S. Transmission Lines Restoration Using Resilience Analysis. In Proceedings of the 2018 Twentieth International Middle East Power Systems Conference (MEPCON), Cairo, Egypt, 18–20 December 2018; pp. 249–253. [CrossRef]

14. Li, Z.; Li, X.; Kang, R. System resilience modeling and assessment based on minimal paths. In Proceedings of the 2016 Prognostics and System Health Management Conference (PHM-Chengdu), Chengdu, China, 19–21 October 2016; pp. 1–4. [CrossRef]

15. Tortós, J.Q.; Terzija, V. A smart power system restoration based on the merger of two different strategies. In Proceedings of the 2012 3rd IEEE PES Innovative Smart Grid Technologies Europe (ISGT Europe), Berlin, Germany, 14–17 October 2012; pp. 1–8.

16. Liu, C.; Wu, M.; Deng, Y. Start-up sequence of generators in power system restoration avoiding the backtracking algorithm. In Proceedings of the IEEE Power and Energy Society General Meeting, Vancouver, BC, Canada, 21–25 July 2013; pp. 1–5. [CrossRef]

17. Zhong, H.; Gu, X.; Zhu, X. Optimization Method for Unit Restarting during Power System Black-Start Restoration. In Proceedings of the 2012 Asia-Pacific Power and Energy Engineering Conference, Shanghai, China, 27–29 March 2012; pp. 1–4. [CrossRef]

18. Rachunok, B.; Nateghi, R. The sensitivity of electric power infrastructure resilience to the spatial distribution of disaster impacts. *Reliab. Eng. Syst. Saf.* **2020**, *193*, 106658. [CrossRef]

19. Arab, A.; Khodaei, A.; Han, Z.; Khator, S.K. Proactive recovery of electric power assets for resiliency enhancement. *IEEE Access* **2015**, *3*, 99–109. [CrossRef]

20. De Caro, F.; Carlini, E.M.; Villacci, D. Flexibility sources for enhancing the resilience of a power grid in presence of severe weather conditions. In Proceedings of the 2019 AEIT International Annual Conference (AEIT), Florence, Italy, 18–20 September 2019; pp. 1–6.

21. Panteli, M.; Pickering, C.; Wilkinson, S.; Dawson, R.; Mancarella, P. Power system resilience to extreme weather: Fragility modeling, probabilistic impact assessment, and adaptation measures. *IEEE Trans. Power Syst.* **2016**, *32*, 3747–3757. [CrossRef]

22. Sahner, R.A.; Trivedi, K.; Puliafito, A. *Performance and Reliability Analysis of Computer Systems: An Example-Based Approach Using the SHARPE Software Package*; Springer Science & Business Media: Berlin, Germany, 2012.

23. Margaritis, D.; Thrun, S. Bayesian network induction via local neighborhoods. In *Advances in Neural Information Processing Systems*; The MIT Press: Boston, MA, USA, 2000; pp. 505–511.

Article

A Modern Data-Mining Approach Based on Genetically Optimized Fuzzy Systems for Interpretable and Accurate Smart-Grid Stability Prediction

Marian B. Gorzałczany *,† , Jakub Piekoszewski † and Filip Rudziński †

Department of Electrical and Computer Engineering, Kielce University of Technology, Al. 1000-lecia P.P. 7, 25-314 Kielce, Poland; j.piekoszewski@tu.kielce.pl (J.P.); f.rudzinski@tu.kielce.pl (F.R.)
* Correspondence: m.b.gorzalczany@tu.kielce.pl
† These authors contributed equally to this work.

Received: 30 March 2020; Accepted: 15 May 2020; Published: 18 May 2020

Abstract: The main objective and contribution of this paper was/is the application of our knowledge-based data-mining approach (a fuzzy rule-based classification system) characterized by a genetically optimized interpretability-accuracy trade-off (by means of multi-objective evolutionary optimization algorithms) for transparent and accurate prediction of decentral smart grid control (DSGC) stability. In particular, we aim at uncovering the hierarchy of influence of particular input attributes upon the DSGC stability. Moreover, we also analyze the effect of possible "overlapping" of some input attributes over the other ones from the DSGC-stability perspective. The recently published and available at the UCI Database Repository *Electrical Grid Stability Simulated Data Set* and its input-aggregate-based concise version were used in our experiments. A comparison with 39 alternative approaches was also performed, demonstrating the advantages of our approach in terms of: (i) interpretable and accurate fuzzy rule-based DSGC-stability prediction and (ii) uncovering the hierarchy of DSGC-system's attribute significance.

Keywords: decentral smart grid control (DSGC); interpretable and accurate DSGC-stability prediction; data mining; computational intelligence; fuzzy rule-based classifiers; multi-objective evolutionary optimization

1. Introduction

The stability of electrical grids depends on the balance between electricity generation and electricity demand. In conventional power systems, such a balance is achieved through demand-driven electricity production. Nowadays, however, due to a gradual shift from fossil-based power generation to renewable energy sources, the grid topologies are becoming more decentralized and the flow of power is becoming more bidirectional [1]. That means that consumers may function as both producers and consumers at the same time; they are often referred to as prosumers [2]. The volatile and fluctuating nature of renewable energy sources poses a significant challenge as far as design strategies and control of electric power grids are concerned. In order to balance the supply and demand in such fluctuating power grids, various smart grid approaches have been proposed. A key idea they implement is to regulate the consumers' demand [3], usually referred to as the demand response strategy [4,5].

The changes in the consumption of electricity (in comparison with normal patterns of consumption) by customers in reaction to the changes in the price of electricity are referred to as demand response. There are two general approaches to defining the electricity price and to communicating it to consumers. A conventional approach—using costly information and communication technology infrastructure [6] is based on extensive communication between producers and consumers [7,8], raising questions, however, of cybersecurity and privacy protection [9,10]. In contrast, an alternative and novel approach referred to as decentral smart grid control (henceforward DSGC) [11] avoids massive communication between prosumers by binding the electricity price to the grid frequency which can be easily measured by means of cheap equipment by particular prosumers. During power excess, the frequency increases, whereas in times of underproduction, it decreases [12]. In that way, DSGC introduces real-time pricing allowing prosumers to easily control their momentary demand on the basis of the grid frequency. For DSGC systems to be successfully applied, they must be able to maintain grid stability for rapid changes in electricity prices and for different levels of reaction times and price sensitivity of particular grid participants [13,14].

Data mining approaches are well suited for decision support in management, control, and stability analysis of power systems including also decentralized smart grids. That is due to the availability of big amounts of simulated data on various aspects of the power systems operation; see, e.g., [15–18]. Many data mining methods have been used in the considered research field—see the next section for a brief review. However, their significant shortcoming is usually their non-transparent, black-box, and accuracy-only-oriented nature. It means that they do not provide any deeper (or any) explanations and justifications of the decisions made. As well, they do not provide any insight into mechanisms governing a given system. A similar remark has been formulated in the most recently published work [19]: "Various machine-learning and data-mining algorithms have been applied to the decentralized management and control of microgrids... Transparency is typically not the priority for most machine learning algorithms" (a quote from [19]). The present work is our attempt to address the smart-grid-stability prediction problem in an effective way by providing a solution coming from the knowledge-based data-mining field and characterized by both high interpretability and transparency and high accuracy.

The main goal and contribution of this work is the application of our knowledge-based data-mining technique, i.e., fuzzy rule-based classifiers (FRBCs) with a genetically optimized interpretability-accuracy trade-off (see, e.g., [20–23] for details) to transparent and accurate prediction of DSGC stability. In particular, we aim at uncovering the hierarchy of influence of particular input attributes upon the DSGC stability. Moreover, we will also analyze the effect of possible "overlapping" of some input attributes over the other ones from the DSGC-stability perspective. In our approach to designing FRBCs from data, measures of FRBCs' interpretability and accuracy are treated as separate performance indices and optimization objectives. Due to their complementary/contradictory nature, we employ multi-objective evolutionary optimization algorithms (MOEOAs) in the process of the FRBCs' structure and parameter optimization which is also equivalent to the FRBC's interpretability-accuracy trade-off optimization (related work is reviewed in [24] and—for the case of single-objective optimization of the system—in [25,26]).

The remaining parts of the paper are organized as follows: We start with a review of related work regarding applications of data mining techniques to various aspects of electricity grid and microgrid operations. Next, the recently published *Electrical Grid Stability Simulated Data Set* available at the UCI Database Repository (https://archive.ics.uci.edu/ml) and its input-aggregate-based concise version proposed in [14] are characterized. Both data sets are used in our experiments. Then, main building blocks of our FRBCs are presented. Next, the FRBCs' learning and optimization process is outlined. Two MOEOAs are independently used and compared; namely, the well-known strength Pareto evolutionary algorithm 2 (SPEA2) [27] and our generalization of SPEA2, referred to as SPEA3 [28–30]. In turn, the previously outlined main goal of the paper, including the application of our approach to the DSGC-stability prediction using two aforementioned simulated data sets and a comparative analysis with as many as 39 alternative approaches, is presented and discussed.

2. Related Work

Various data-mining and machine-learning models and algorithms have been applied to security, stability prediction/monitoring, management, and control of electricity grids and microgrids.

An approach using an artificial neural network (ANN for short) for generating security boundaries and their visualization to transmission system operators within a power system in California is presented in [15]. The resulting security boundaries are visualized in the form of the so-called nomograms (the total number of simulations performed is equal to 1792). Extreme learning machines (ELMs for short)—a special class of ANNs—are used in [31] to improve the online learning speed and parameter tuning of the real-time transient-stability assessment model for earlier detection of the risk of blackouts. ANNs are also used to construct new monitoring methods for smart power grids. Such a method and a virtual test evaluating its performance are presented in [32]. A multilayer ANN with four hidden layers trained using a deep reinforcement learning algorithm is applied to a smart grid optimization task in [33]. In turn, a contextual anomaly detection ANN-based approach for cyber-physical security in smart grids is presented in [34]. The simulation experiments show that the contextual anomaly detection performs over 55% better than the point anomaly detection.

Support vector machines (SVMs for short) and ANNs supported by some feature selection methods are applied in [17] to the analysis of the transient stability of a large-scale Brazilian power system (the data are generated in 1242 simulation runs). SVMs and random forests (RFs for short) are used in [35] to detect smart grid devices compromised by cyber attacks. The proposed framework in different evaluation scenarios yields high accuracy (91% on average) which confirms its effectiveness at overcoming the compromised smart grid devices problem. Core vector machines (CVMs for short)—faster and big-data-oriented extensions of SVMs—are used in [36] for an online transient stability assessment of a power system by mapping that problem as a two-class classification task. An online approach makes it attractive to be used in real time applications. Genetic-algorithm-based SVMs (GA-SVMs for short) are used (and compared with conventional SVMs and ANNs) in [37] for an online voltage-stability monitoring and prediction. Their effectiveness is demonstrated by applying them to the New England 39-bus system and to the real Indian Northern Region Power Grid system.

Decision trees (DTs for short) are used (and compared with ANNs and SVMs) in [18] for prediction of the transient instability of a large-scale Iranian national grid following the test on a small 9-bus system. DTs are also used in [14] to classify the DSGC stability conditions based on the response of heterogeneous consumers to provide some insight into the relationship between the input space parameters and the grid stability.

As far as other selected data-mining techniques are concerned, the so-called active learning solution is applied in [38] to the voltage-stability prediction problem. The active learning approach interacts with the online prediction and offline training process to enhance the well-known data-mining methods (DTs, ANNs, SVMs, RFs, and radial basis function networks). In turn, modeling a non-linear security boundary by (i) using features formed as monomials of the original input up to a certain level and (ii) using kernel ridge regression to solve the problem of a large number of features is proposed in [16]. The potential of the proposed method is demonstrated by simulating the aforementioned New England 39-bus system and a larger power system with 470 buses. Next, the study [39] presents a cooperative multi-agent approach for solving the complex problems of energy management in a stand-alone solar microgrid using fuzzy logic systems and a reinforcement learning method referred to as fuzzy Q-learning. Moreover, the study [19] applies an optimized data-matching algorithm referred to as transparent open box learning network to the DSGC stability prediction. This study demonstrates the importance of compound feature selection in the considered stability prediction problem. The overwhelming majority of the above-listed studies presents the accuracy-oriented approaches. The transparency is not their priority and they usually do not provide an insight into the considered systems.

3. *Electrical Grid Stability Simulated Data Set* and Its Input-Aggregate-Based Version for Stability Prediction of a Four-Node Star System Implementing the DSGC Concept

As already mentioned in the Introduction of this paper, the *Electrical Grid Stability Simulated Data Set* available since November 2018 at the UCI Database Repository (https://archive.ics.uci.edu/ml) is the first set used in our experiments. This data set is the outcome of a simulation experiment using a two-part mathematical model of a four-node star grid implementing the DSGC concept. The first part of the model describes the physical dynamics of electric power generation and its connection with consumption loads. The second part is an economic structure which binds the electricity price to the grid frequency (see [11,13,14,19] for details). In the simulation experiment, three key input variables of the model were selected (each allowed to vary independently) for each of the four grid participants. These key input variables include: (i) P_j, $j = 1, 2, 3, 4$ (referred to as p1, ..., p4 in the simulated data set) describing the mechanical power produced (for $j = 1$) or consumed (for $j = 2, 3, 4$); (ii) τ_j, $j = 1, 2, 3, 4$ (referred to as tau1, ..., tau4 in the simulated data set) describing reaction time of each grid participant to an electricity price change; and (iii) γ_j, $j = 1, 2, 3, 4$ (referred to as g1, ..., g4 in the simulated data set) which is a coefficient proportional to price elasticity for each grid participant. Figure 1 illustrates the structure of the DSGC system considered in the simulation experiment and the feasible solution space values (boundary conditions) for all the previously listed key input variables. Except for P_1 ($P_1 = -(P_2 + P_3 + P_4)$ as shown in Figure 1), they all are sampled as uniform distributions throughout their respective feasible spaces to initialize and launch the simulations (10000 simulation runs were performed). Several other input variables of the two-part mathematical model of the DSGC system are kept unchanged during the simulation process. They include: (i) the averaging time T required to measure the price signal ($T = 2s$), (ii) the coupling strength K proportional to line capacity ($K = 8s^{-2}$), and (iii) the damping constant α ($\alpha = 0.1s^{-1}$). The model's output variable (referred to as stab in the simulated data set), i.e., the stability metric (ranging from -0.0808 to $+0.1094$), is quantified such that a negative value of that metric indicates that the grid is stable, whereas a positive value indicates that the grid is unstable. In the simulated data set the stab-variable is accompanied by a stabf label of the grid stability—a categorical attribute taking values from a two-element set: "stable" for stab < 0 and "unstable" for stab > 0. The details regarding the simulation experiment are presented in [14].

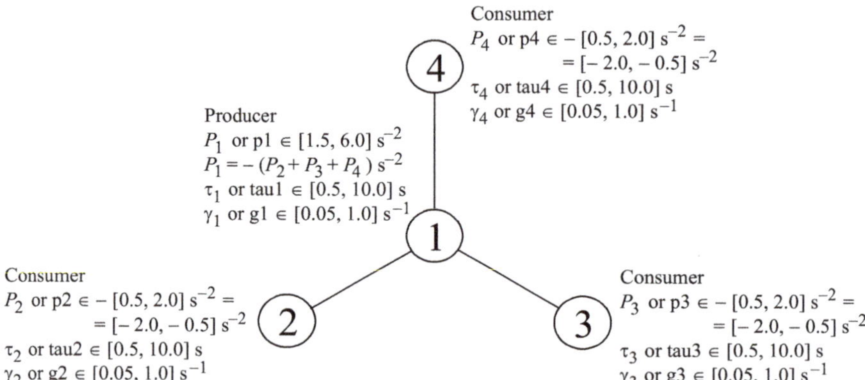

Figure 1. An illustration of the four-node star DSGC system structure in a simulation experiment.

Therefore, the characteristics of the *Electrical Grid Stability Simulated Data Set* collecting the simulation experiment results and used in our experiments is the following. It contains 10, 000 records (instances). Each record is characterized by 12 input attributes—tau1, tau2, tau3, tau4, p1, p2, p3, p4, g1, g2, g3, and g4; and two output attributes—stab and stabf, out of which only the stabf labels are used in our experiments (see Table 1 for details).

Table 1. Details of particular records of the *Electrical Grid Stability Simulated Data Set* used in our experiments.

No.	Attribute Name	Attribute Type	Attribute Domain Details		min.	max.	avg.	std. dev.
1.	tau1	numerical	reaction time of electricity producer (in sec.)		0.50079	9.9995	5.25	2.7424
2.	tau2	numerical	⎫		0.50014	9.9998	5.25	2.7424
3.	tau3	numerical	⎬ reaction time of electricity consumers (in sec.)		0.50079	9.9995	5.25	2.7424
4.	tau4	numerical	⎭		0.50047	9.9994	5.25	2.7424
5.	p1	numerical	nominal power produced (positive values)		1.5826	5.8644	3.57	0.75212
6.	p2	numerical	⎫		−1.9999	−0.50011	−1.25	0.43301
7.	p3	numerical	⎬ nominal power consumed (negative values)		−1.9999	−0.50007	−1.25	0.43301
8.	p4	numerical	⎭		−1.9999	−0.50002	−1.25	0.43301
9.	g1	numerical	gamma coefficient [1] for electricity producer		0.050009	0.99994	0.525	0.27424
10.	g2	numerical	⎫		0.050053	0.99994	0.525	0.27424
11.	g3	numerical	⎬ gamma coefficient [1] for electricity consumers		0.050054	0.99998	0.525	0.27424
12.	g4	numerical	⎭		0.050028	0.99993	0.525	0.27424
13.	stab [2]	numerical	the maximal real part of the characteristic equation root (if positive—the system is linearly unstable)		−0.08076	0.1094	0.01573	0.03691
14.	stabf	categorical (class label)	2 class labels; class balance: "nstable" class (63.8% of all records) and "stable" class (36.2% of all records)					

Attributes 1–12 are *input attributes*; attributes 13–14 are *output attr.*

[1] coefficient proportional to price elasticity; [2] attribute "stab" is not used in our experiments.

A more concise representation of the above-characterized simulated data set is proposed in [14]. It is based on aggregates—such as minimum, maximum, and average (mean) values—of input attributes across all grid participants. For instance, for the reaction time τ_j, $j = 1, \ldots, 4$, the input aggregates are the following: $\tau_{min} = \min_{j=1,\ldots,4} \tau_j$, $\tau_{avg} = \frac{1}{4} \sum_{j=1}^{4} \tau_j$, and $\tau_{max} = \max_{j=1,\ldots,4} \tau_j$. In the modified simulated data set, they are referred to as tau_min, tau_avg, and tau_max; analogously—p_min, p_avg, and p_max for P_j and g_min, g_avg, and g_max for γ_j, $j = 1, \ldots, 4$ (see Table 2 for details). The modified data set (referred to as the *Concise Simulated Data Set*) is the second set used in our experiments.

Table 2. Details of particular records of the input-aggregate-based version of the *Electrical Grid Stability Simulated Data Set* (referred to as the *Concise Simulated Data Set*) used in our experiment.

No.	Attribute Name	Attribute Type	Attribute Domain Details		min.	max.	avg.	std. dev.
1.	tau_min	numerical	⎫		0.50014	8.874	2.4011	1.5649
2.	tau_avg	numerical	⎬ min., avg., and max. reaction time of		0.87644	9.446	5.25	1.3743
3.	tau_max	numerical	⎭ electricity producer/consumers (in sec.)		1.0217	9.9998	8.0944	1.5482
4.	p_min	numerical	⎫		0.5644	1.9999	1.6254	0.2913
5.	p_avg	numerical	⎬ min., avg., and max. nominal power		0.5275	1.9548	1.25	0.2507
6.	p_max	numerical	⎭ consumed		1.5526	5.8644	3.75	0.7521
7.	g_min	numerical	⎫		0.05	0.8884	0.2397	0.1552
8.	g_avg	numerical	⎬ min., avg., and max. gamma coefficient [1]		0.1194	0.9536	0.525	0.1369
9.	g_max	numerical	⎭ for electricity producer/consumers		0.1459	0.9999	0.811	0.1539
10.	stabf	categorical (class label)	2 class labels; class balance: "unstable" class (63.8% of all records) and "stable" class (36.2% of all records)					

Attributes 1–9 are *input attributes*; attribute 10 is *out. attr.*

[1] coefficient proportional to price elasticity.

4. Methodology: Main Components of the Proposed FRBCs Designed from Data Using MOEOAs

In this section we briefly present the main components of our approach to designing FRBCs from data by means of MOEOAs. They are used to perform FRBCs' learning and structure-and-parameter optimization, which also results in the optimization of FRBCs' interpretability-accuracy trade-off (see [20–22,28] for details and discussion). The following components are characterized: learning data set, representation of input attributes and class labels, FRBC knowledge base, objectives of the FRBCs' evolutionary optimization, original dedicated genetic operators introduced, and MOEOAs used in our experiments. An FRBC with n input attributes $x_1, x_2, \ldots, x_n$ and an output—a fuzzy set over the set $Y = \{y_1, y_2, \ldots, y_c\}$ of c class labels—is considered. Our approach can process both numerical and categorical attributes. However, in the DSGC-stability prediction problem, only numerical attributes occur. Hence, only numerical attributes will be considered from now on.

Learning data set L: The construction of the proposed FRBC is based on the data set L which contains K input–output samples:

$$L = \{x_k^{(lrn)}, y_k^{(lrn)}\}_{k=1}^{K}, \tag{1}$$

where $x_k^{(lrn)} = (x_{1k}^{(lrn)}, x_{2k}^{(lrn)}, \ldots, x_{nk}^{(lrn)}) \in X = X_1 \times X_2 \times \cdots \times X_n$ ($\times$ stands for Cartesian product of ordinary sets) represents the collection of input attributes and $y_k^{(lrn)}$ represents the corresponding class label ($y_k^{(lrn)} \in Y$) for the k-th data sample, $k = 1, 2, \ldots, K$.

Representation of input attributes and class labels: Each numerical attribute x_i, $i \in \{1, 2, \ldots, n\}$ is represented by a_i fuzzy sets $A_{ik_i} \in F(X_i)$, $k_i = 1, 2, \ldots, a_i$, where $F(X_i)$ is a family of all fuzzy sets defined in the universe X_i, $i = 1, 2, \ldots, n$. A_{i1} represents an S-type fuzzy set (corresponding to linguistic term "*Small*"), A_{ia_i} represents an L-type set (corresponding to linguistic term "*Large*"), and $A_{i2}, A_{i3}, \ldots, A_{i,a_i-1}$ represent M-type sets (corresponding to linguistic terms "*Medium 1,*' "*Medium 2,*' ... , "*Medium $a_i - 2$*"). For simplicity, A_{ik_i}s denote the corresponding linguistic terms also. Figure 2 shows trapezoidal membership functions for S, M, and L-type fuzzy sets used in our experiments. In turn, each class label y_j, $j \in \{1, 2, \ldots, c\}$ is represented by a fuzzy singleton $B_j = B_j^{(singl.)}$ with the following membership function: $\mu_{B_j^{(singl.)}}(y) = 1$ for $y = y_j$ and 0 elsewhere.

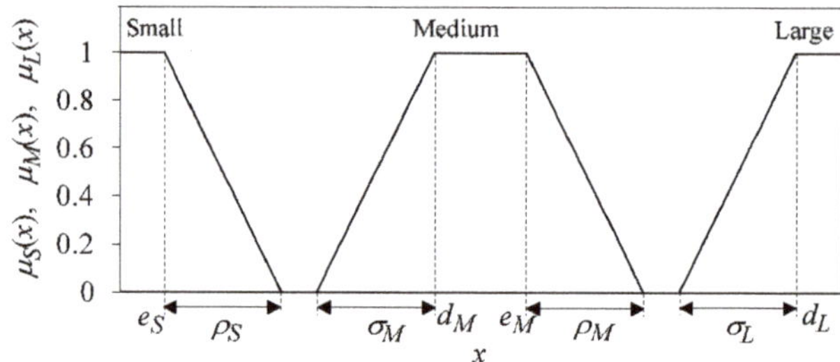

Figure 2. S-type, M-type, and L-type fuzzy sets with trapezoidal membership functions and their parameters.

FRBC's knowledge base contains R genetically optimized fuzzy rules discovered in the learning data set (1). The form of the r-th rule, $r = 1, 2, \ldots, R$, is the following (the overall number of rules R changes as the learning progresses):

$$\textbf{IF } [x_1 \text{ is } A_{1,sw_1^{(r)}}]_{(sw_1^{(r)}>0)} \textbf{ AND...AND } [x_n \text{ is } A_{n,sw_n^{(r)}}]_{(sw_n^{(r)}>0)} \textbf{ THEN } y \text{ is } B_{j^{(r)}}^{(singl.)}. \tag{2}$$

Formula $[expression]_{(condition)}$ in (2) represents conditional inclusion of the $[expression]$-part into a given rule assuming that the $(condition)$-part is fulfilled. In turn, $sw_i^{(r)}$ denotes a switch-parameter which controls the presence/absence of the i-th input attribute in the r-th rule, $i = 1, 2, \ldots, n$.

$sw_i^{(r)} \in \{0, 1, 2, \ldots, a_i\}$, where a_i is the number of fuzzy sets (and the corresponding linguistic terms) defined for the i-th attribute. For $sw_i^{(r)} = 0$, the i-th attribute is removed from (not active in) that rule, whereas for $sw_i^{(r)} > 0$ the component $[x_i \text{ is } A_{ik_i}]$ ($k_i = sw_i^{(r)}$) is included (active) in that rule.

An FRBC's knowledge base contains two separate modules; i.e., a rule-base-structure module RB and a data-base module DB. We propose a simple, direct, and thus computationally efficient RB representation as follows:

$$RB = \{sw_1^{(r)}, sw_2^{(r)}, \ldots, sw_n^{(r)}, j^{(r)}\}_{r=1}^{R}.$$ (3)

In turn, DB contains tunable and non-tunable parts. The tunable part contains parameters of membership functions of fuzzy sets representing particular numerical input attributes. These parameters—subject to tuning during the FRBCs' learning and MOEOA-based optimization—include (see Figure 2): e_S and ρ_S for S-type fuzzy sets; σ_M, d_M, e_M, and ρ_M for M-type fuzzy sets; and σ_L and d_L for L-type fuzzy sets. The non-tunable part of DB contains the set of class labels $Y = \{y_1, y_2, \ldots, y_c\}$.

Objectives of the FRBC's evolutionary optimization: Two separate optimization objectives are considered; i.e., the accuracy and the interpretability of the system. The FRBC's accuracy measure (subject to maximization) is defined as follows [20–22,40]:

$$Q_{ACC}^{(lrn)} = 1 - Q_{RMSE}^{(lrn)},$$ (4)

where

$$Q_{RMSE}^{(lrn)} = \sqrt{\frac{1}{Kc} \sum_{k=1}^{K} \sum_{j=1}^{c} \left[\mu_{B_k^{(singl.)(lrn)}}(y_j) - \mu_{B_k'}(y_j) \right]^2}.$$ (5)

$\mu_{B_k'}(y)$ is the membership function of fuzzy set B_k', which is a response of system (2) for the learning data sample $x_k^{(lrn)}$. In turn, $\mu_{B_k^{(singl.)(lrn)}}(y)$ (equal to 1 for $y = y_k^{(lrn)}$ and to 0 elsewhere) is the membership function of fuzzy singleton $B_k^{(singl.)(lrn)}$, which is the desired fuzzy-singleton response for that sample ($Q_{RMSE}^{(lrn)} \in [0, 1]$).

As far as the FRBC's interpretability is concerned, we use the notion of interpretability in a broader sense, which includes two essential aspects; i.e., the FRBC's complexity and semantic aspects of the FRBC's operation. We evaluate the FRBC's complexity-related interpretability using the following measure (subject to maximization):

$$Q_{INT} = 1 - Q_{CPLX},$$ (6)

where

$$Q_{CPLX} = \frac{Q_{RINP} + Q_{INP} + Q_{FS}}{3},$$ (7)

and

$$Q_{RINP} = \frac{1}{R} \sum_{r=1}^{R} \frac{n_{INP}^{(r)} - 1}{n - 1}, \quad Q_{INP} = \frac{n_{INP} - 1}{n - 1}, \quad Q_{FS} = \frac{n_{FS} - 1}{\sum_{i=1}^{n} a_i - 1}, \quad n > 1.$$ (8)

The FRBC's complexity measure Q_{CPLX} (7) takes its values from interval [0, 1], where 0 represents the minimal complexity and 1 the maximal one. Q_{CPLX} is an average of three sub-measures that evaluate: (i) an average complexity of particular rules Q_{RINP} (8) ($n_{INP}^{(r)}$ in (8) denotes the number of active input attributes in the r-th rule); (ii) the complexity of the whole system in terms of its active inputs Q_{INP} (8) (n_{INP} in (8) denotes the number of active inputs in the whole system); and (iii) the whole-system complexity in terms of its active fuzzy sets Q_{FS} (8) (n_{FS} in (8) denotes the numbers of active fuzzy sets (linguistic terms) in the whole system).

The FRBC's semantics-related interpretability is addressed by us by imposing—as optimization constraints—the so-called strong fuzzy partitions (SFPs) [41] upon domains of particular numerical attributes. SFPs are special class of fuzzy partitions; namely, for any domain value, the sum of the values of all membership functions constituting SFP is equal to 1. It can be shown [41] that SFPs satisfy the desired demands regarding the semantics-related interpretability. Straightforward implementation of SFP requirements for the case of trapezoidal membership functions can be formulated in the following way (see Figure 3 for illustration of the SFP-requirements' implementation for the three-set SFP of x_i-domain):

$$\sigma_{ik_i} = \rho_{i,k_i-1} = d_{ik_i} - e_{i,k_i-1}, \quad k_i = 2,3,\ldots,a_i \tag{9}$$

and, obviously,

$$e_{i1} \leq d_{i2} \leq e_{i2} \leq \cdots \leq d_{i,a_i-1} \leq e_{i,a_i-1} \leq d_{ia_i}, \; i = 1,2,\ldots,n. \tag{10}$$

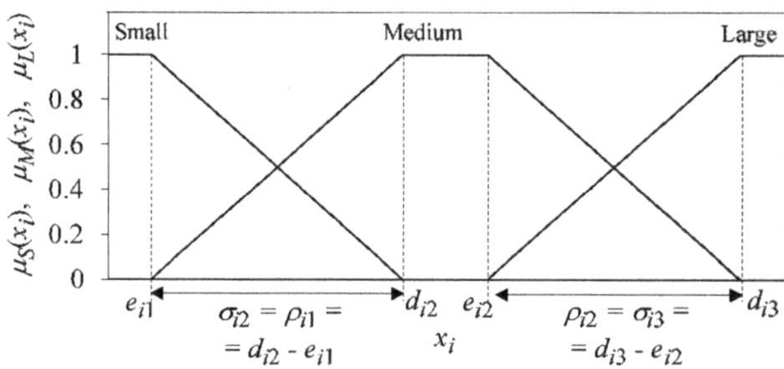

Figure 3. Illustration of the strong fuzzy partition (SFP) requirements for the three-set SFP of x_i-domain.

Original, dedicated genetic operators introduced: A population of FRBC's knowledge bases (each represented by its RB and DB) is processed during the genetic learning process. We developed original crossover and mutation operators for the transformation of the RB population. The crossover operator processes two individuals (i.e., two RBs) containing R_1 and R_2 fuzzy rules, respectively, by performing one sub-operation randomly selected from the set of five sub-operations referred to as Cro-RB1, Cro-RB2, ..., Cro-RB5. They are defined as follows:

Cro-RB1 (labeled as "exchange of many fuzzy rules") operates in two stages. First, for the r-th rule in both RBs, $r = 1,2,\ldots,min(R_1,R_2)$, a random-switch condition (equivalent to the random selection of 1 from the set $\{0,1\}$) is checked. Then, the r-th rules from both RBs are exchanged, provided that the random-switch condition is fulfilled. Second, each of the remaining rules of the larger RB is moved to the smaller RB, provided that its random-switch condition is fulfilled.

Cro-RB2 (labeled as "exchange of a single fuzzy rule") is similar to Cro-RB1 but the Cro-RB1 activities are carried out unconditionally and only once for randomly selected single rule from the larger RB.

Cro-RB3 (labeled as "exchange of many fuzzy sets in many fuzzy rules") is analogous to the first stage of Cro-RB1. This time, the random-switch condition is run for each input attribute and for the output class label from the corresponding rules coming from both RBs. Fuzzy sets describing a given input attribute or class label in both RBs are exchanged provided that their random-switch conditions are fulfilled.

Cro-RB4 (labeled as "exchange of many fuzzy sets in a single fuzzy rule") performs the activities of Cro-RB3 unconditionally and only once for a randomly selected single rule from the first RB and its counterpart from the second RB.

Cro-RB5 (labeled as "exchange of a single fuzzy set") is a special case of Cro-RB4. The activities of Cro-RB4 are performed unconditionally and only once for a randomly selected input attribute or output class label.

The mutation operator for the RB transformation processes a single individual (i.e., a single RB) by performing one sub-operation randomly selected from the set of four sub-operations referred to as Mut-RB1, Mut-RB2, Mut-RB3, Mut-RB4. They are defined as follows:

Mut-RB1 (labeled as "fuzzy rule insertion") inserts into RB a new fuzzy rule (2) with randomly selected values of switches $sw_i^{(r)}$ ($sw_i^{(r)} \in \{0, 1, 2, \ldots, a_i\}$, $i = 1, 2, \ldots, n$) and class label $j^{(r)}$ ($j^{(r)} \in \{1, 2, \ldots, c\}$).

Mut-RB2 (labeled as "fuzzy rule deletion") removes a randomly selected fuzzy rule from the RB.

Mut-RB3 (labeled as "change a single fuzzy set") randomly selects one fuzzy rule from the RB and its one i-th input attribute or output class label j. Next, it randomly selects a new value of its switch sw_i or class label j.

Mut-RB4 (labeled as "change of an input in a fuzzy rule") randomly selects: (i) one fuzzy rule from the RB, (ii) its one active input attribute (i.e., with $sw_{i_1} > 0$), and (iii) its one non-active input attribute (i.e., with $sw_{i_2} = 0$). Then, the first attribute is set to be non-active (i.e., $sw_{i_1} = 0$) and the second attribute—to be active (i.e., $sw_{i_2} > 0$) in that rule.

The crossover and mutation operators for the RB-population transformation are followed by three RB-repairing operators. The first operator removes "empty" fuzzy rules, i.e., the rules with all non-active input attributes (with $sw_i = 0$ for $i = 1, 2, \ldots, n$), whereas the second one removes rule duplicates. In turn, the third operator adds—for each class label that is currently not represented in RB—one fuzzy rule with that class label (in order to preserve the principle "at least one fuzzy rule per class in RB").

DB population transformation is performed by means of separate genetic operators. The crossover operator (processing two individuals; i.e., two DBs), randomly selects one fuzzy set from each DB. New values of d and e-parameters, which characterize membership functions of both fuzzy sets (see Figure 2) are calculated as linear combinations of their old values from both sets; they also must fulfill condition (10). New values of σ and ρ-parameters are calculated from (9) using new values of d and e-parameters. The mutation operator (processing a single individual, i.e., a single DB) randomly selects one fuzzy set from the DB and one of its two parameters d and e (say, d is selected). Its new value $d_{new} = d + rand(-0.2, 0.2)[x_{i,max} - x_{i,min}]$, where $rand(\cdot)$ returns a random number from the assumed interval and $[x_{i,min}, x_{i,max}]$ is a range of the domain of the selected set. New values of σ and ρ-parameters are calculated from (9).

MOEOAs used in our experiments: The performances of different MOEOAs are usually evaluated and compared in terms of three aspects [42]. First, the accuracy of generated non-dominated solutions is considered. The accuracy represents the closeness of those solutions to either the Pareto-optimal solutions, or if they are not available, to reference solutions. Second, the spread of the solution set is investigated; i.e., how well these solutions arrive at the extrema of the Pareto-optimal or reference solution sets. The spread is usually represented by the distance between the extreme solutions in the set. Third, the distribution of the solutions in the set, i.e., how evenly distributed they are along the approximation of Pareto-optimal front in the objective space, is considered. A set of solutions which are more accurate and are characterized by a higher spread and a better-balanced distribution outperforms the alternative solution sets.

For the purpose of comparison, two MOEOAs are used in experiments reported in this work; i.e., the well-known SPEA2 method and our generalization of SPEA2, referred to as SPEA3. In comparison with SPEA2, the proposed SPEA3 approach generates sets of non-dominated solutions characterized by higher spread and a more even distribution in the objective space. The essence of the proposed SPEA2's generalization consists of replacing its environmental selection procedure with our original algorithm, which improves both the spread and distribution balance of generated solutions. The environmental selection consists of selecting a representation of the best solutions from all solutions obtained so far and keeping them in an external archive of a fixed size. In SPEA2 all non-dominated solutions from the archive and the current population are copied to the next-generation archive (to fill the archive, the best dominated solutions are copied to it). In the case of overfilling the archive, a truncation procedure is run to reduce the archive size to the predefined level. Thus, only the truncation procedure (if activated) contributes to improving the distribution balance and diversity of the final set of solutions. On the contrary, the proposed environmental selection algorithm implemented in SPEA3 is fully-oriented towards achieving these goals. The archive is iteratively increased by gradual adding of carefully selected non-dominated solutions from the current population (in each subsequent generation, a single solution characterized by possible longest and similar distances to its near neighbors is added). Then, a process of relocation of particular archived solutions aiming at increasing average distances between solutions and their nearest neighbors is carried out. Said relocation is performed by gradual replacement of archived solutions by new solutions selected from the population in such a way as (i) to maximize the distances between extreme solutions (which results in improving the spread of solutions) and (ii) to minimize the distance differences between neighboring solutions (which results in improving the distribution of solutions). Concluding, in our approach, the complementary operations of increasing and reducing the archive aim at obtaining the best available distribution balance and spread of solutions belonging to Pareto-front approximation (see [28,29] for detailed presentation and discussion).

5. Experiments (Application of Our Approach to DSGC-Stability Prediction) and Discussion

In this section we present the results of two experiments (for both data sets characterized in Section 3 of this paper) regarding the application of our approach to interpretable and accurate DSGC-stability prediction.

5.1. Application to Electrical Grid Stability Simulated Data Set

We start with revealing some details of the operation of our method for constructing FRBCs from the considered simulated data set. Figure 4a presents two 10-element-collections of non-dominated solutions (optimized FRBCs) generated in a single run of our FRBCs' design technique employing, independently, our SPEA3 and SPEA2 algorithms. Both experiments of genetic learning have been performed for a learning-test data split with 1:9 ratio. It means that only 10% of the whole original data set (preserving the class proportions) is used as the learning data to construct the system, whereas the remaining 90% of the original data are used for the system's testing. It is worth emphasizing that such a data split poses a significant challenge to any system's design technique. Each collection of solutions of Figure 4a represents the best available approximations of Pareto-optimal solutions generated either by our SPEA3 or SPEA2. Particular solutions from a given collection (front) are characterized by different levels of optimized interpretability-accuracy trade-off. Thus, the user can select a single solution (a specific FRBC) with a desired degree of compromise between the interpretability and accuracy.

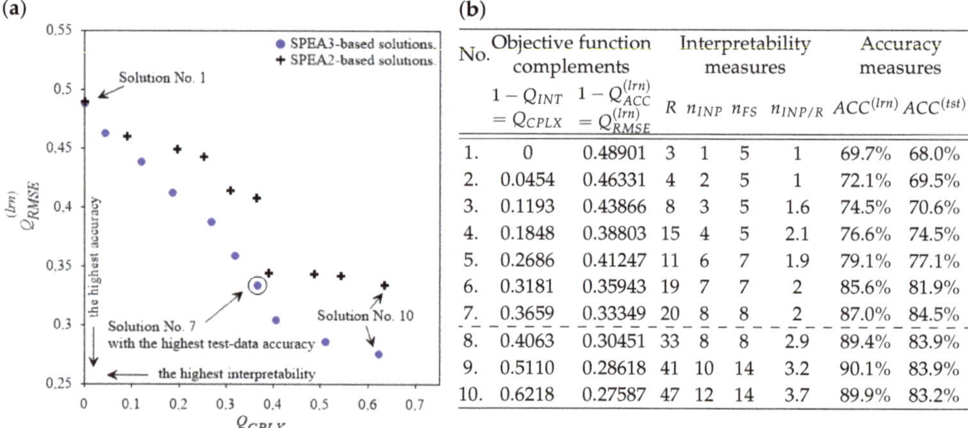

(b)

No.	Objective function complements		Interpretability measures				Accuracy measures	
	$1 - Q_{INT}$ $= Q_{CPLX}$	$1 - Q_{ACC}^{(lrn)}$ $= Q_{RMSE}^{(lrn)}$	R	n_{INP}	n_{FS}	$n_{INP/R}$	$ACC^{(lrn)}$	$ACC^{(tst)}$
1.	0	0.48901	3	1	5	1	69.7%	68.0%
2.	0.0454	0.46331	4	2	5	1	72.1%	69.5%
3.	0.1193	0.43866	8	3	5	1.6	74.5%	70.6%
4.	0.1848	0.38803	15	4	5	2.1	76.6%	74.5%
5.	0.2686	0.41247	11	6	7	1.9	79.1%	77.1%
6.	0.3181	0.35943	19	7	7	2	85.6%	81.9%
7.	0.3659	0.33349	20	8	8	2	87.0%	84.5%
8.	0.4063	0.30451	33	8	8	2.9	89.4%	83.9%
9.	0.5110	0.28618	41	10	14	3.2	90.1%	83.9%
10.	0.6218	0.27587	47	12	14	3.7	89.9%	83.2%

Figure 4. (a) The best Pareto-front approximation generated by our SPEA3 and SPEA2; (b) interpretability and accuracy measures of SPEA3-based solutions from (a) (*Electrical Grid Stability Simulated Data Set*).

Figure 4a shows that our SPEA3-based approach generates the collection of solutions characterized by: (i) a much-better-balanced distribution in the objective space (i.e., the solutions are distributed along the front in a much more even way) and (ii) higher accuracy (i.e., the closeness to Pareto-optimal front). Therefore, our SPEA3-based approach outperforms its SPEA2-based counterpart (see Section 4 of this paper for MOEOAs' performance evaluation criteria). The interpretability and accuracy-related numerical details of all SPEA3-based solutions (FRBCs) from Figure 4a are collected in Figure 4b, in which $n_{INP/R}$ is the number of input attributes per rule, whereas $ACC^{(lrn)}$ is the percentage of correct decisions in the learning set and $ACC^{(tst)}$—in the test set (the remaining parameters were defined in Section 4 of the paper).

Table 3. Fuzzy rule bases for SPEA3-based solutions (FRBCs) 1–3 from Figure 4.

No.	**Fuzzy Classification Rules**
Solution No. 1 ($ACC^{(tst)} = 68\%$):	
1.	**IF** tau1 is Small **THEN** stable
2.	**IF** tau1 is Medium **THEN** unstable
3.	**IF** tau1 is Large **THEN** unstable
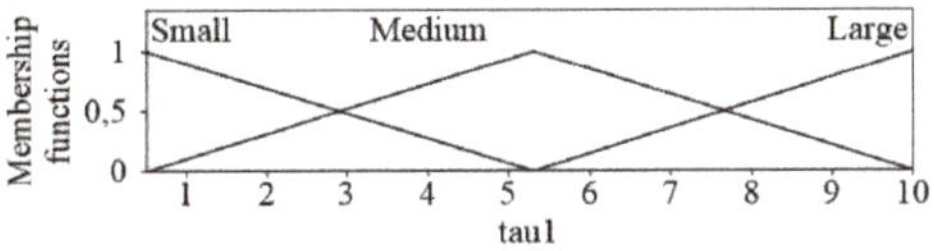	
Solution No. 2 ($ACC^{(tst)} = 69.5\%$):	
1-3.	*These rules are the same as rules* 1-3 *from Solution No.* 1.
4.	**IF** tau4 is Small **THEN** stable
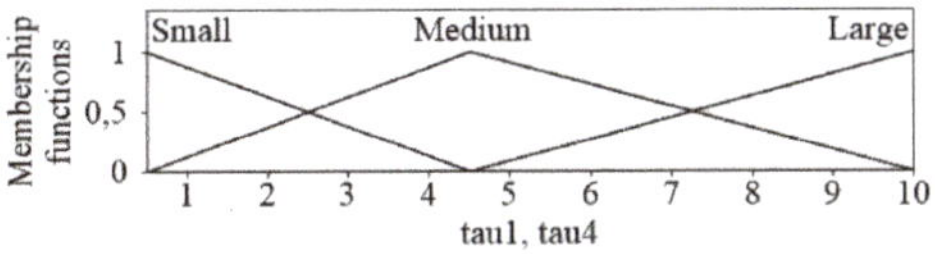	
Solution No. 3 ($ACC^{(tst)} = 70.6\%$):	
1,2.	*These rules are the same as rules* 1, 2 *from Solution No.* 2.
3.	**IF** tau1 is Large **AND** tau4 is Medium **THEN** unstable
4.	*This rule is the same as rule* 4 *from Solution No.* 2.
5.	**IF** tau1 is Large **AND** tau4 is Large **THEN** unstable
6.	**IF** tau3 is Small **AND** tau4 is Medium **THEN** stable
7.	**IF** tau3 is Medium **AND** tau4 is Medium **THEN** unstable
8.	**IF** tau3 is Large **AND** tau4 is Large **THEN** unstable
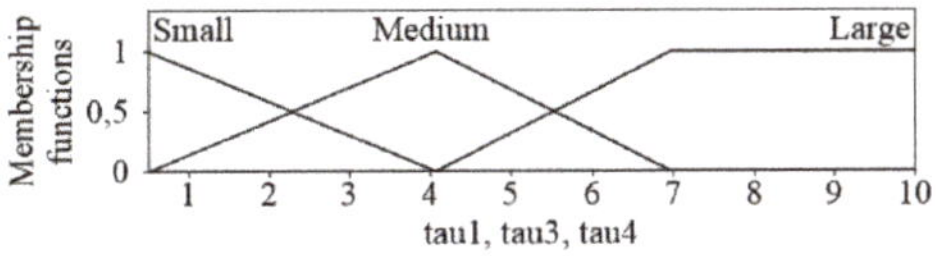	

Table 4. Fuzzy rule base for SPEA3-based solution (FRBC) 7 from Figure 4 which is characterized by the highest test-data accuracy.

No.	Fuzzy Classification Rules
Solution No. 7 ($ACC^{(tst)} = 84.5\%$):	
1.	**IF** tau1 is Small **AND** tau2 is Small **THEN** stable
2.	**IF** tau1 is Small **AND** tau3 is Small **THEN** stable
3.	**IF** tau1 is Small **AND** tau4 is Small **THEN** stable
4.	**IF** tau1 is Small **AND** g2 is Small **THEN** stable
5.	**IF** tau1 is Small **AND** g4 is Small **THEN** stable
6.	**IF** tau1 is Medium **AND** g4 is Large **THEN** unstable
7.	**IF** tau1 is Large **AND** g1 is Large **THEN** unstable
8.	**IF** tau2 is Small **AND** tau3 is Small **THEN** stable
9.	**IF** tau2 is Small **AND** tau4 is Small **THEN** stable
10.	**IF** tau2 is Small **AND** g3 is Small **THEN** stable
11.	**IF** tau2 is Medium **AND** g2 is Large **THEN** unstable
12.	**IF** tau2 is Large **AND** g2 is Large **THEN** unstable
13.	**IF** tau3 is Small **AND** tau4 is Small **THEN** stable
14.	**IF** tau3 is Small **AND** g4 is Small **THEN** stable
15.	**IF** tau3 is Large **AND** g3 is Large **THEN** unstable
16.	**IF** tau4 is Small **AND** g3 is Small **THEN** stable
17.	**IF** tau4 is Large **AND** g4 is Large **THEN** unstable
18.	**IF** g1 is Small **AND** g2 is Small **THEN** stable
19.	**IF** g1 is Small **AND** g2 is Medium **AND** g4 is Medium **THEN** stable
20.	**IF** g2 is Small **AND** g3 is Small **THEN** stable

Fuzzy rule bases of exemplary SPEA3-based solutions (FRBCs) are presented in Tables 3 and 4 (membership functions of fuzzy sets representing input attributes used are also included). Table 4 presents the fuzzy rule base for the SPEA3-based solution (FRBC) 7 from Figure 4 which is characterized by the highest test-data accuracy. In turn, Table 3 reveals an interesting regularity, i.e., the fuzzy rule base of the solution 2 contains three rules from the solution 1. In turn, the fuzzy rule base of the solution 3 contains three rules from the solution 2, etc. Therefore, if higher system accuracy is required, then our approach adds some additional fuzzy rules (or extends the earlier-discovered rules) to provide a more detailed, and thus, more accurate description of the considered prediction problem. Said regularity confirms the internal integrity of our approach. This regularity is also illustrated in Table 5 (see, first, Part A of Table 5) in which each black square denotes the presence of a given input attribute in the fuzzy rule base of a given solution (FRBC). $ACC_1^{(tst)}$ is the test-data accuracy of the system exclusively based on the most significant input attribute. In Part A of Table 5 the most significant attribute is tau1, occurring in the solution 1 and giving $ACC_1^{(tst)} = 68\%$. Moreover, $\Delta ACC_j^{(tst)}, j = 2, 3, \ldots$ is the accuracy increase following the inclusion of the 2nd, 3rd, . . . most significant attributes into the system. In Part A of Table 5, the inclusion of tau4 (2nd most significant attribute) yields 1.5% increase in the test-data-accuracy and is related to the solution 2. In turn, the inclusion of tau3 (3rd most significant attribute) gives a further 1.1% test-data-accuracy-increase and is related to the solution 3, etc.

In such a way, we can uncover the hierarchy of significance of particular attributes. However, the obtained attribute-significance hierarchy is correct provided that no "overlapping" of some input attributes over the other ones occurs in the simulated data set. In order to verify that, we remove from the original data set the so-far most significant attribute, i.e., tau1, and we repeat, in an analogous way, the learning experiment. Its results are presented in Part B of Table 5, giving tau4 attribute the most significance place. Tau4 occupied second position in the experiment of Part A. Therefore, we conclude that tau4 is not "overlapped" by tau1. In the next step, we remove tau4 from the present data set and repeat the learning experiment—see Part C of Table 5—obtaining tau3 as the most significant attribute at this stage. It occupied second position in experiment of Part B. Therefore, tau3 is not "overlapped" by tau4. We repeat analogous experiments several times; i.e., we remove the most significant—at a given stage—input attribute and repeat the learning process on the reduced data set—see Parts D–H of Table 5. In such a way, we arrive to the real final hierarchy of input attribute significance in the DSGC-stability prediction problem. It is shown in the left part of Table 6 that $ACC_1^{(tst)}$ has the same meaning as in Table 5; i.e., it is the test-data-accuracy of the system exclusively based on a single attribute listed to the left of $ACC_1^{(tst)}$ in Table 6.

The results of an alternative approach addressing the hierarchy of importance of particular input attributes in the considered simulated data set and proposed in [43] are presented in the right part of Table 6. The following remarks are formulated in [43]: "The time response of the consumers and producer to the price fluctuation play a more important role in the grid stability, as compared to the price elasticity index... Unstable grid conditions generally prevail for higher values of τ and γ... The power consumption pattern does not show any difference under stable and unstable grid conditions thus conforming with the analysis of feature selection obtained." (a quote from [43]). The results of our experiments are consistent with all the above comments.

Table 5. Illustration of attribute presence and significance in the *Electrical Grid Stability Simulated Data Set*.

Attribute Name	$ACC_1^{(tst)}$ $\Delta ACC_j^{(tst)}$ $j=2,3,\ldots$	Attribute Presence in the Rules of Solution No.:									
		1	2	3	4	5	6	7	8	9	10
Part A—Number of input attributes: 12 (all)											
tau1	68.0%	■	■	■	■	■	■	■	■	■	■
tau4	+1.5%		■	■	■	■	■	■	■	■	■
tau3	+1.1%			■	■	■	■	■	■	■	■
tau2	+3.9%				■	■	■	■	■	■	■
g3	} +2.6%					■	■	■	■	■	■
g4						■	■	■	■	■	■
g2	+4.8%						■	■	■	■	■
g1	+2.6%							■	■	■	■
p1	} −0.4%									■	■
p4										■	■
p2	} −0.7%										■
p3											■

(Left margin label: Attribute significance — Low ← → High)

Table 5. Cont.

Attribute Name	$ACC_1^{(tst)}$ / $\Delta ACC_j^{(tst)}$, $j=2,3,\dots$	1	2	3	4	5	6	7	8	9	10
Part B—Number of input attributes: 11											
tau1	removed attribute										
tau4	67.7%	■	■	■	■	■	■	■	■	■	■
tau3	+0.7%		■	■	■	■	■	■	■	■	■
tau2	+2.0%			■	■	■	■	■	■	■	■
g3	+2.2%				■	■	■	■	■	■	■
g4	+3.4%					■	■	■	■	■	■
g2	+3.9%						■	■	■	■	■
g1	+1.8%								■	■	■
p2	} −0.6%									■	■
p4										■	■
p3	} +0.2%										■
p1											■
Part C—Number of input attributes: 10											
tau1	} removed attributes										
tau4											
tau3	66.1%	■	■	■	■	■	■	■	■	■	■
tau2	+2.2%		■	■	■	■	■	■	■	■	■
g3	+2.4%				■	■	■	■	■	■	■
g2	+3.0%					■	■	■	■	■	■
g4	+2.5%						■	■	■	■	■
g1	+3.7%							■	■	■	■
p1	+0.3%								■	■	■
p3	−0.2%									■	■
p2	} −0.1%										■
p4											■
Part D—Number of input attributes: 9											
tau1	} removed attributes										
tau4											
tau3											
tau2	66%	■	■	■	■	■	■	■	■	■	■
g4	+2.6%		■	■	■	■	■	■	■	■	■
g3	+1.7%				■	■	■	■	■	■	■
g2	+2.8%						■	■	■	■	■
g1	+1.4%							■	■	■	■
p1	−0.3%								■	■	■
p2	} −0.8%									■	■
p3										■	■
p4	−2.4%										■
Part E—Number of input attributes: 8											
tau1	} removed attributes										
tau4											
tau3											
tau2											
g3	64%	■	■	■	■	■	■	■	■	■	■
g2	+3.6%		■	■	■	■	■	■	■	■	■
g1	+0.6%				■	■	■	■	■	■	■
g4	+2.2%					■	■	■	■	■	■
p3	−0.1%							■	■	■	■
p2	−0.5%								■	■	■
p1	0.0%									■	■
p4	−0.2%										■

The "Attribute Presence in the Rules of Solution No.:" heading spans columns 1–10. The left margin of each part is labelled "Attribute significance" with an arrow from *Low* (bottom) to *High* (top).

Table 5. *Cont.*

The left margin of the table carries the vertical label "Attribute significance", with an arrow from *Low* ↓ (bottom) to *High* ↑ (top).

Attribute Name	$ACC_1^{(tst)}$ / $\Delta ACC_j^{(tst)}$, $j = 2, 3, \ldots$	Attribute Presence in the Rules of Solution No.:									
		1	**2**	**3**	**4**	**5**	**6**	**7**	**8**	**9**	**10**
Part F—Number of input attributes: 7											
tau1											
tau4	removed attributes										
tau3											
tau2											
g3											
g2	+62.7%	■	■	■	■	■	■	■	■	■	■
g4	+5.1%		■	■	■	■	■	■	■	■	■
p1	−3.1%					■	■	■	■	■	■
g1	+3.1%						■	■	■	■	■
p4	−1.9%								■	■	■
p3	+0.6%									■	■
p2	+0.8%										■
Part G—Number of input attributes: 6											
tau1											
tau4	removed attributes										
tau3											
tau2											
g3											
g2											
g4	62.2%	■	■	■	■	■	■	■	■	■	■
p1	+1.3%		■	■	■	■	■	■	■	■	■
g1	+1.6%					■	■	■	■	■	■
p2	+0.1%								■	■	■
p4	+0.3%									■	■
p3	−0.6%										■
Part H—Number of input attributes: 5											
tau1											
tau4											
tau3	removed attributes										
tau2											
g3											
g2											
g4											
g1	62.1%	■	■	■	■	■	■	■	■	■	■
p1	+1.1%		■	■	■	■	■	■	■	■	■
p2	0.0%						■	■	■	■	■
p4	−0.1%								■	■	■
p3	−0.1%									■	■

Table 6. Final hierarchy of attribute significance—a comparison of our approach and an alternative method of [43] (*Electrical Grid Stability Simulated Data Set*).

	Our Approach		Alternative Approach of [43] (Panda and Das (2019))	
	Attribute Name	$ACC_1^{(tst)}$	Attribute Name	Importance
High ↑	tau1	68.2%	tau2	0.064
	tau4	67.8%	tau4	0.061
	tau3	66%	tau3	0.06
	tau2	66%	tau1	0.058
	g3	64%	g3	0.057
	g2	62.7%	g2	0.05
Low ↓	g4	62.2%	g4	0.044
	g1	62.1%	g1	0.041
	p1, p2, p3, and p4[1]	-	p1, p2, p3, and p4	≪ 0.001

(Leftmost column label, rotated: *Attribute significance*.)

[1] The attributes p1, p2, p3, and p4 do not occur in the fuzzy rule base of our most accurate solution 7 (see Table 4) and have not been tested.

The cross-validation experiment with 1:9 learning-test data split ratio is the next and important part of our work. Each single learning experiment begins with generation of a Pareto-front approximation. Next, a single solution with, first, the highest test-data accuracy, and second, the highest interpretability, is chosen from that front approximation. In turn, the results from all partial experiments are averaged. The experiment is then repeated 10 times for different initializations of our approach. The averaged results are shown in the last row of Table 7, which also collects the results of as many as 39 alternative approaches applied to the considered data set; i.e., the *Electrical Grid Stability Simulated Data Set*. There are four groups of them. The first one, considered in [44], includes four methods: logistic regression (LR), random forest (RF), gradient boosted trees (GBT), and multilayer perceptron classifier (MPC). Each of them was supported, independently, by three feature selection algorithms: multivariate adaptive regression splines (MARS) algorithm, binary kangaroo mob optimization feature selection (BKMOFS) algorithm, and binary particle swarm optimization feature selection (BPSOFS) algorithm. The second group, considered in [45], includes seven methods: *k*-nearest neighbors (kNN) algorithm, support vector machine (SVM), radial basis function (RBF) network, decision trees, RF, naive Bayes approach, and quadratic discriminant analysis (QDA). The third group, considered in [46], includes eight methods: fine and bagged-tree algorithms, fine and weighted-kNN, LR, and linear, quadratic, and cubic-SVMs. The last group, considered in [47], includes six methods: the stochastic damped regularized LBFGS algorithm (Sd-REG-LBFGS; LBFGS stands for the limited memory Broyden–Fletcher–Goldfarb–Shanno algorithm), stochastic damped regularized LBFGS without regularization (SdLBFGS), robust stochastic approximation (RSA), stochastic approximation averaging (SAA), stochastic gradient difference (SGD), and a method of gradient-based stochastic optimization (Adam [48]). Each of them was used, independently, with logistic regression and Bayesian logistic regression algorithms.

The overwhelming majority of the alternative approaches of Table 7 are the black-box methods focused only on the system's accuracy. Moreover, for methods applied in [44–46] only the learning-data-accuracy is available. Therefore, it is not possible to assess their generalizing capabilities (usually measured by test-data-accuracy), which belong to essential performance-evaluation criteria of any system designed from data. Some of the methods of [44] operate on more than 100 features, whereas the remaining approaches typically use 12 attributes; i.e., all the input attributes from the simulated data set. The experiments of [47] are performed for the 4:1 learning-test data split ratio, which significantly favors them in regard to our experiments using only 1:9 learning-test data split ratio. Despite that, our approach provides not only comparable test-data accuracy but also a detailed insight—in the form of fuzzy linguistic rules—into the mechanisms governing the DSGC stability.

Table 7. Results of our approach and comparison with alternative methods for the *Electrical Grid Stability Simulated Data Set*.

Source	Method		Learn-to-Test Ratio	Number of Runs	Average Accuracy Measures for Learning and Test Data		Average Interpretability Measures			
					$ACC^{(lrn)}$	$ACC^{(tst)}$	$\bar{R}$	$\bar{n}_{ATR}$	$\bar{n}_{FS}$	$\bar{n}_{ATR/R}$
[44] (Moldovan and Salomie (2019))	LR	with MARS-based feature selection algorithm	-	n/a	63.9%	n/a	-	10	-	-
	RF		-	n/a	87.8%	n/a	n/a	10	n/a	n/a
	GBT		-	n/a	88%	n/a	n/a	10	n/a	n/a
	MPC		-	n/a	91.9%	n/a	-	10	-	-
	LR	with BKMOFS-based feature selection algorithm	-	n/a	66.3%	n/a	-	102	-	-
	RF		-	n/a	83.7%	n/a	n/a	102	n/a	n/a
	GBT		-	n/a	88.8%	n/a	n/a	102	n/a	n/a
	MPC		-	n/a	91.5%	n/a	-	102	-	-
	LR	with BPSOFS-based feature selection algorithm	-	n/a	64.7%	n/a	-	111	-	-
	RF		-	n/a	84.3%	n/a	n/a	111	n/a	n/a
	GBT		-	n/a	87%	n/a	n/a	111	n/a	n/a
	MPC		-	n/a	93.8%	n/a	-	111	-	-
[45] (Karim (2019))	kNN		n/a	n/a	80.4%	n/a	-	12	-	-
	SVM		n/a	n/a	82.2%	n/a	-	12	-	-
	RBF		n/a	n/a	62.7%	n/a	-	12	-	-
	Decision Tree		n/a	n/a	84%	n/a	n/a	12	n/a	n/a
	RF		n/a	n/a	88.5%	n/a	n/a	12	n/a	n/a
	Naive Bayes		n/a	n/a	83.6%	n/a	-	12	-	-
	QDA		n/a	n/a	82.7%	n/a	-	12	-	-
[46] (Balali et al. (2020))	Fine Tree		n/a	n/a	84.2%	n/a	n/a	12	n/a	n/a
	Bagged Tree		n/a	n/a	91.5%	n/a	n/a	12	n/a	n/a
	Fine KNN		n/a	n/a	81.1%	n/a	-	12	-	-
	Weighted KNN		n/a	n/a	86.6%	n/a	-	12	-	-
	LR		n/a	n/a	81.5%	n/a	-	12	-	-
	Linear SVM		n/a	n/a	81.5%	n/a	-	12	-	-
	Quadratic SVM		n/a	n/a	94.2%	n/a	-	12	-	-
	Cubic SVM		n/a	n/a	96.9%	n/a	-	12	-	-
[47] (Chen, Chan, Wu and Lam (2019))	Sd-REG-LBFGS	with logistic regression	4:1	50	n/a	87.55%	-	12	-	-
	SdLBFGS		4:1	50	n/a	87.68%	-	12	-	-
	RSA		4:1	50	n/a	86.72%	-	12	-	-
	SAA		4:1	50	n/a	86.72%	-	12	-	-
	SGD		4:1	50	n/a	86.72%	-	12	-	-
	Adam		4:1	50	n/a	52.69%	-	12	-	-
	Sd-REG-LBFGS	with Bayesian logistic regression	4:1	50	n/a	87.56%	-	12	-	-
	SdLBFGS		4:1	50	n/a	87.67%	-	12	-	-
	RSA		4:1	50	n/a	87.27%	-	12	-	-
	SAA		4:1	50	n/a	87.27%	-	12	-	-
	SGD		4:1	50	n/a	87.27%	-	12	-	-
	Adam		4:1	50	n/a	64.53%	-	12	-	-
This paper	Our approach based on SPEA3		1:9	10	91.1%	85.5%	22.7	9.7	26.9	4.1

n/a stands for not available.

5.2. Application to an Input-Aggregate-Based, Concise Version of the Simulated Data Set of Section 5.1

As already said in Section 3 of this paper, a concise, input-aggregate-based representation of the *Electrical Grid Stability Simulated Data Set* was proposed in [14]. According to [14]: "Since the system has symmetries, we hypothesize that a more concise representation of simulation results is feasible based on input aggregates, i.e., features. To create features, we take the minimum, maximum and mean values across all N participants of each input; e.g., $min\tau_j$ for $j = 1, \ldots, N$." (a quote from [14]).

Following that, we constructed such a set (referred to as the *Concise Simulated Data Set*) as shown in Section 3. It will be used in experiments reported in this section.

Figure 5a presents a 10-element-collection of non-dominated solutions (optimized FRBCs) generated in a single run of our SPEA3 algorithm. Similarly to Section 5.1, the genetic learning experiment has been performed for the learning-test data split with a 1:9 ratio. Figure 5b presents the interpretability and accuracy-related numerical details for solutions of Figure 5a.

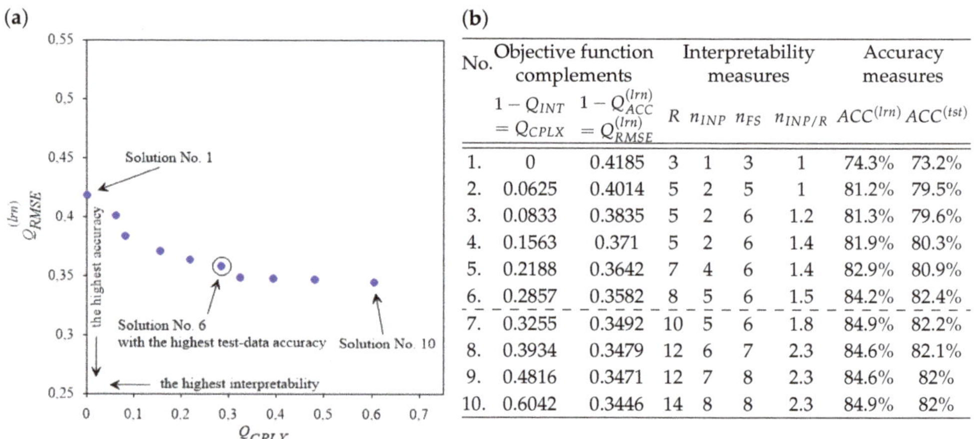

(a)

(b)

No.	Objective function complements		Interpretability measures				Accuracy measures	
	$1 - Q_{INT}$ $= Q_{CPLX}$	$1 - Q_{ACC}^{(lrn)}$ $= Q_{RMSE}^{(lrn)}$	R	n_{INP}	n_{FS}	$n_{INP/R}$	$ACC^{(lrn)}$	$ACC^{(tst)}$
1.	0	0.4185	3	1	3	1	74.3%	73.2%
2.	0.0625	0.4014	5	2	5	1	81.2%	79.5%
3.	0.0833	0.3835	5	2	6	1.2	81.3%	79.6%
4.	0.1563	0.371	5	2	6	1.4	81.9%	80.3%
5.	0.2188	0.3642	7	4	6	1.4	82.9%	80.9%
6.	0.2857	0.3582	8	5	6	1.5	84.2%	82.4%
7.	0.3255	0.3492	10	5	6	1.8	84.9%	82.2%
8.	0.3934	0.3479	12	6	7	2.3	84.6%	82.1%
9.	0.4816	0.3471	12	7	8	2.3	84.6%	82%
10.	0.6042	0.3446	14	8	8	2.3	84.9%	82%

Figure 5. (**a**) The best Pareto-front approximation generated by our SPEA3; (**b**) interpretability and accuracy measures of solutions from (**a**) (*Concise Simulated Data Set*).

Table 8—presenting fuzzy rule bases of exemplary solutions from Figure 5—shows the same regularity as for experiments of Section 5.1. Namely, if the higher accuracy of the system is required, our approach adds additional rules or extends the existing ones to provide a more detailed and accurate model for decision support. Table 9 presents fuzzy rule base of the solution 6 which achieves the highest test-data accuracy (see Figure 5). Transformation of fuzzy classification rules from Table 9 into decision-tree form is shown in Figure 6. It reveals the mechanisms governing the DSGC-stability from the perspective of essential input aggregates such as tau_min, tau_avg, tau_max, g_avg, and g_max.

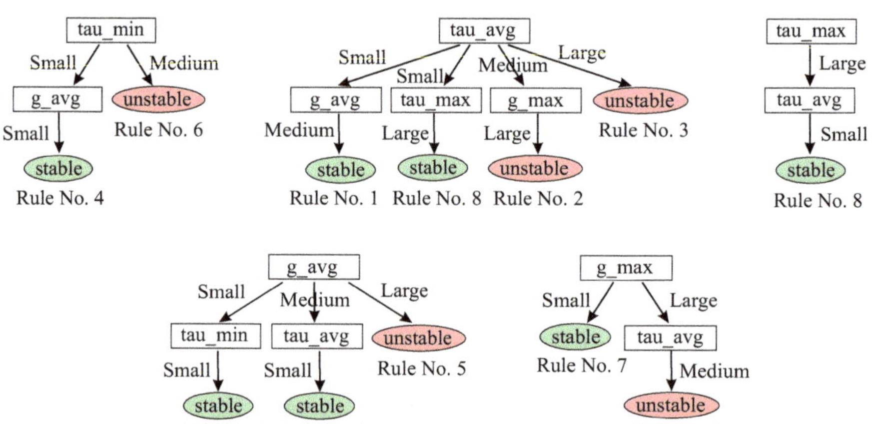

Figure 6. Transformation of fuzzy classification rules from Table 9 into decision-tree form.

Table 8. Fuzzy rule bases for SPEA3-based solutions (FRBCs) 1–3 from Figure 5.

No.	Fuzzy Classification Rules

Solution No. 1 ($ACC^{(tst)} = 73.2\%$):

1.	**IF**	tau_avg is Small **THEN** stable
2.	**IF**	tau_avg is Medium **THEN** unstable
3.	**IF**	tau_avg is Large **THEN** unstable

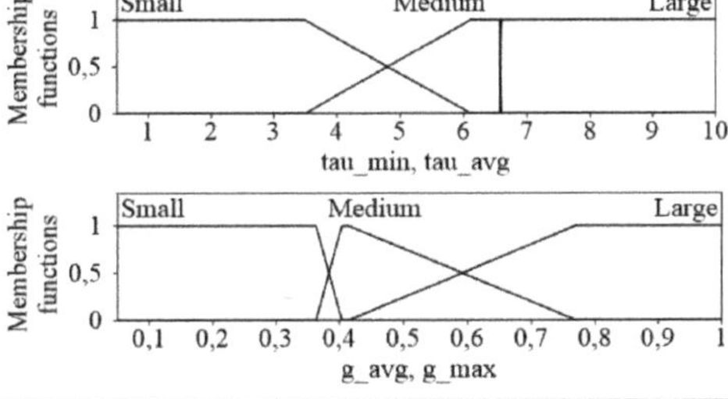

Solution No. 2 ($ACC^{(tst)} = 79.5\%$):

1-3.	*These rules are the same as rules 1-3 from Solution No. 1.*	
4.	**IF**	g_avg is Small **THEN** stable
5.	**IF**	g_avg is Large **THEN** unstable

Membership functions of fuzzy sets for particular input attributes are the same as in Solution No. 5

Solution No. 3 ($ACC^{(tst)} = 79.6\%$):

1.	**IF**	tau_avg is Small **AND** g_avg is Medium **THEN** stable
2-5.	*These rules are the same as rules 2-5 from Solution No. 2.*	

Membership functions of fuzzy sets for particular input attributes are the same as in Solution No. 5

Solution No. 4 ($ACC^{(tst)} = 80.3\%$):

1.	*This rule is the same as rule 1 from Solution No. 3.*	
2.	**IF**	tau_avg is Medium **AND** g_max is Large **THEN** unstable
3-5.	*These rules are the same as rules 3-5 from Solution No. 3.*	

Membership functions of fuzzy sets for particular input attributes are the same as in Solution No. 5

Solution No. 5 ($ACC^{(tst)} = 80.9\%$):

1-3.	*These rules are the same as rules 1-3 from Solution No. 4.*	
4.	**IF**	tau_min is Small **AND** g_avg is Small **THEN** stable
5.	*This rule is the same as rule 5 from Solution No. 4.*	
6.	**IF**	tau_min is Medium **THEN** unstable
7.	**IF**	g_max is Small **THEN** stable

Table 9. Fuzzy rule base for SPEA3-based solution (FRBC) 6 from Figure 5.

No.		Fuzzy Classification Rules
Solution No. 6 ($ACC^{(tst)} = 82.4\%$):		
1.	**IF**	tau_avg is Small **AND** g_avg is Medium **THEN** stable
2.	**IF**	tau_avg is Medium **AND** g_max is Large **THEN** unstable
3.	**IF**	tau_avg is Large **THEN** unstable
4.	**IF**	tau_min is Small **AND** g_avg is Small **THEN** stable
5.	**IF**	g_avg is Large **THEN** unstable
6.	**IF**	tau_min is Medium **THEN** unstable
7.	**IF**	g_max is Small **THEN** stable
8.	**IF**	tau_avg is Small **AND** tau_max is Large **THEN** stable

Rules 1-7 are the same as rules 1-7 from Solution No. 5.
(for the convenience of the reader, they have been repeated here)

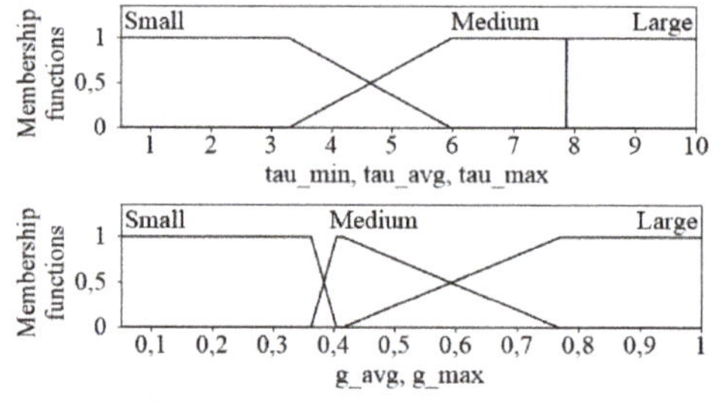

In the work [14] we can say that, "If $min(\tau_j) < 2.1$ and $avg(\gamma_j) \geq 0.5$ and $avg(\tau_j) < 4.8$ and $max(\tau_j) \geq 8$ then the system is stable. This means that, in a stable grid, a consumer may have a reaction time higher than $\tau_c \approx 8s$) as long as there is a consumer reacting quite fast, and the average reaction time is moderate" (a quote from [14]). We can formulate a similar conclusion based on the much simpler (and thus more interpretable) rule 8 from our fuzzy rule base of Table 9. It is obvious that as long as tau_avg is small, then tau_min is also small; therefore, it is not necessary to include tau_min into the rule. In turn, the grid stability suffers when all participants react relatively slowly (see the rule 6 of Table 9) or very slowly (see the rule 3 of that table). In general, higher values of tau and g lead to unstable grid conditions—see the rules 2 and 5 of Table 9 from the "unstable"-class perspective and the rules 1, 4, and 7 of that table—from the "stable"-class point of view.

In the final part of this section, we would like to put a "bridge" between solutions for the *Electrical Grid Stability Simulated Data Set* of Section 5.1 and its input-aggregate-based *Concise Simulated Data Set* of Section 5.2. Figure 7 shows the best Pareto-front approximations generated by our SPEA3 algorithm for both considered data sets (they were earlier presented independently in Figures 4a and 5a). Figure 7 confirms an intuitively obvious regularity; namely, the solutions for the input-aggregate-based concise data set are more interpretability-oriented ones, whereas the solutions for the original data set with non-aggregated input attributes are more accuracy-oriented ones.

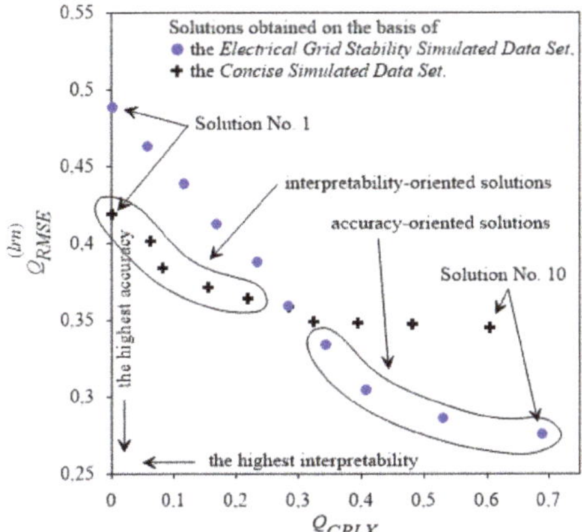

Figure 7. The best Pareto-front approximations generated by our SPEA3 for the *Electrical Grid Stability Simulated Data Set* and the *Concise Simulated Data Set*.

6. Conclusions

In this paper we address the problem of transparent and accurate prediction of decentral smart grid control stability using our knowledge-based data-mining approach implemented as the fuzzy rule-based classifier. Our approach employs multi-objective evolutionary optimization algorithms to optimize the interpretability-accuracy trade-off of the classification system. The transparency and interpretability (i.e., the ability to provide the user with understandable and compact explanations of generated predictions) and the accuracy (i.e., the ability to generate correct and precise predictions) are important aspects of the operation of decision support systems for smart grid stability prediction. Compact, linguistic, fuzzy classification rules generated by our approach—due to their high readability and easy-to-grasp interpretation—belong to the most effective knowledge-representation structures in the considered domain. Our approach, in a single run, generates a set of non-dominated solutions (a collection of fuzzy classification systems) characterized by different levels of optimized interpretability-accuracy trade-off; the user can select a single solution according to his/her needs. Recently published and available at the UCI Database Repository (https://archive.ics.uci.edu/ml) *Electrical Grid Stability Simulated Data Set* and its input-aggregate-based concise version referred to as *Concise Simulated Data Set* are used in our experiments.

The contribution of this paper is twofold. First, by means of broad cross-validation-based experiments, we show that our approach significantly outperforms alternative methods (altogether 39 alternative methods are considered) in terms of the transparency and interpretability of generated predictions while remaining competitive or superior in terms of the accuracy of those predictions. It is worth emphasizing that the overwhelming majority of the existing studies on smart grid stability prediction concentrate on the accuracy-oriented approaches not providing an insight into the prediction mechanisms.

Second, our approach—besides being interpretable and accurate in the considered domain—is also an effective method for uncovering the hierarchy of significance of particular input attributes contributing to the smart grid stability prediction process. In order to uncover the real attribute-significance hierarchy we also analyze the possible "overlapping" of some input attributes over the other ones from the DSGC-stability perspective.

Our further work will concentrate on improving the optimization of the systems' interpretability-accuracy trade-off. It is essential from the point of view of generating highly interpretable and highly accurate modern systems (cf. explainable artificial intelligence [49,50] or interpretable machine learning [51,52] systems) for decision support in various areas of applications including smart grid modeling and control.

Author Contributions: Conceptualization, M.B.G., J.P. and F.R.; formal analysis, M.B.G., J.P. and F.R.; investigation, M.B.G., J.P. and F.R.; methodology, M.B.G., J.P. and F.R.; writing–original draft, M.B.G., J.P. and F.R.; writing–review and editing, M.B.G., J.P. and F.R. All authors have read and agreed to the published version of the manuscript.

Funding: This research received no external funding.

Conflicts of Interest: The authors declare no conflict of interest.

References

1. Ackermann, T.; Andersson, G.; Söder, L. Distributed generation: A definition. *Electr. Power Syst. Res.* **2001**, *57*, 195–204. [CrossRef]
2. Kotler, P. The prosumer movement: A new challenge for marketers. *Adv. Consum. Res.* **1986**, *13*, 510–513.
3. Butler, D. Energy efficiency: Super savers: Meters to manage the future. *Nature* **2007**, *445*, 586–588. [CrossRef]
4. Albadi, M.; El-Saadany, E. A summary of demand response in electricity markets. *Electr. Power Syst. Res.* **2008**, *78*, 1989–1996. [CrossRef]
5. Palensky, P.; Dietrich, D. Demand side management: Demand response, intelligent energy systems, and smart loads. *IEEE Trans. Ind. Inform.* **2011**, *7*, 381–388. [CrossRef]
6. Ernst & Young GmbH. *Cost-Benefit Analysis for the Comprehensive Use of Smart Metering Systems—Final Report—Summary*; On behalf of the Federal Ministry of Economics and Technology, Germany, 2013. Available online: https://www.bmwi.de/Redaktion/EN/Publikationen/cost-benefit-analysis-for-the-comprehensive-use-of-smart-metering-systems.html (accessed on 1 May 2020).
7. Kok, K.; Warmer, C.; Kamphuis, R. PowerMatcher: multiagent control in the electricity infrastructure. In Proceedings of the Fourth International Joint Conference on Autonomous Agents and Multiagent Systems, Utrecht, The Netherlands, 25–29 July 2005; pp. 75–82.
8. Gungor, V.C.; Sahin, D.; Kocak, T.; Ergut, S.; Buccella, C.; Cecati, C.; Hancke, G.P. Smart grid technologies: Communication technologies and standards. *IEEE Trans. Ind. Inform.* **2011**, *7*, 529–539. [CrossRef]
9. Ericsson, G.N. Cyber security and power system communication—essential parts of a smart grid infrastructure. *IEEE Trans. Power Deliv.* **2010**, *25*, 1501–1507. [CrossRef]
10. Liu, J.; Xiao, Y.; Li, S.; Liang, W.; Chen, C. Cyber security and privacy issues in smart grids. *IEEE Commun. Surv. Tutor.* **2012**, *14*, 981–997. [CrossRef]
11. Schäfer, B.; Matthiae, M.; Timme, M.; Witthaut, D. Decentral smart grid control. *New J. Phys.* **2015**, *17*, 1–15. [CrossRef]
12. Short, J.; Infield, D.; Freris, L. Stabilization of grid frequency through dynamic demand control. *IEEE Trans. Power Syst.* **2007**, *22*, 1284–1293. [CrossRef]
13. Schäfer, B.; Grabow, C.; Auer, S.; Kurths, J.; Witthaut, D.; Timme, M. Taming instabilities in power grid networks by decentralized control. *Eur. Phys. J. Spec. Top.* **2015**, *225*, 1–19. [CrossRef]
14. Arzamasov, V.; Böhm, K.; Jochem, P. Towards concise models of grid stability. In Proceedings of the 2018 IEEE International Conference on Communications, Control, and Computing Technologies for Smart Grids, SmartGridComm 2018, Aalborg, Denmark, 29–31 October 2018; pp. 1–6.
15. McCalley, J.; Wang, S.; Zhao, Q.; Zhou, G.; Treinen, R.T.; Papalexopoulos, A.D. Security boundary visualization for systems operation. *IEEE Trans. Power Syst.* **1997**, *12*, 940–947. [CrossRef]
16. Jayasekara, B.; Annakkage, U.D. Derivation of an accurate polynomial representation of the transient stability boundary. *IEEE Trans. Power Syst.* **2006**, *21*, 1856–1863. [CrossRef]
17. Moulin, L.S.; da Silva, A.; El-Sharkawi, M.A.; Marks, R.J., II. Support vector machines for transient stability analysis of large-scale power systems. *IEEE Trans. Power Syst.* **2004**, *19*, 818–825. [CrossRef]
18. Amraee, T.; Ranjbar, S. Transient instability prediction using decision tree technique. *IEEE Trans. Power Syst.* **2013**, *28*, 3028–3037. [CrossRef]
19. Wood, D. Predicting stability of a decentralized power grid linking electricity price formulation to grid frequency applying an optimized data-matching learning network to simulated data. *Technol. Econ. Smart Grids Sustain. Energy* **2020**, *5*, 1–21. [CrossRef]
20. Rudziński, F. A multi-objective genetic optimization of interpretability-oriented fuzzy rule-based classifiers. *Appl. Soft Comput.* **2016**, *38*, 118–133. [CrossRef]

21. Gorzałczany, M.B.; Rudziński, F. A multi-objective genetic optimization for fast, fuzzy rule-based credit classification with balanced accuracy and interpretability. *Appl. Soft Comput.* **2016**, *40*, 206–220. [CrossRef]

22. Gorzałczany, M.B.; Rudziński, F. Interpretable and accurate medical data classification—a multi-objective genetic-fuzzy optimization approach. *Expert Syst. Appl.* **2017**, *71*, 26–39. [CrossRef]

23. Gorzałczany, M.B.; Rudziński, F. Handling fuzzy systems' accuracy-interpretability trade-off by means of multi-objective evolutionary optimization methods—selected problems. *Bull. Pol. Acad. Sci. Tech. Sci.* **2015**, *63*, 791–798. [CrossRef]

24. Fazzolari, M.; Alcala, R.; Nojima, Y.; Ishibuchi, H.; Herrera, F. A review of the application of multiobjective evolutionary fuzzy systems: Current status and further directions. *IEEE Trans. Fuzzy Syst.* **2013**, *21*, 45–65. [CrossRef]

25. Gorzałczany, M.B.; Rudziński, F. Accuracy vs. interpretability of fuzzy rule-based classifiers—An evolutionary approach. In *Artificial Intelligence and Soft Computing—ICAISC 2012*; Lecture Notes in Computer Science; Rutkowski, L., Korytkowski, M., Scherer, R., Tadeusiewicz, R., Zadeh, L.A., Żurada, J.M., Eds.; Springer: Berlin, Germany, 2012; Volume 7269, pp. 222–230.

26. Gorzałczany, M.B.; Rudziński, F. A modified Pittsburg approach to design a genetic fuzzy rule-based classifier from data. In *Artificial Intelligence and Soft Computing— ICAISC 2010*; Lecture Notes in Computer Science; Rutkowski, L., Scherer, R., Tadeusiewicz, R., Zadeh, L.A., Żurada, J.M., Eds.; Springer: Berlin, Germany, 2010; Volume 6113, pp. 88–96.

27. Zitzler, E.; Laumanns, M.; Thiele, L. Zitzler, E.; Laumanns, M.; Thiele, L. SPEA2: Improving the strength Pareto evolutionary algorithm for multi-objective optimization. In *Proceedings of the Evolutionary Methods for Design, Optimization and Control with Applications to Industrial Problems (EUROGEN 2001)*; Athens, Greece, 19–21 September 2001, pp. 95–100.

28. Rudziński, F. Finding sets of non-dominated solutions with high spread and well-balanced distribution using generalized strength Pareto evolutionary algorithm. In Proceedings of the 2015 Conference International Fuzzy Systems Association and European Society for Fuzzy Logic and Technology (IFSA-EUSFLAT-15), Gijón, Spain, 30 June–3 July 2015; Atlantis Press: Gijón, Spain, 2015; Volume 89, pp. 178–185.

29. Gorzałczany, M.B.; Rudziński, F. An improved multi-objective evolutionary optimization of data-mining-based fuzzy decision support systems. In Proceedings of the 2016 IEEE International Conference Fuzzy Systems (FUZZ-IEEE), Vancouver, BC, Canada, 25–29 July 2016; pp. 2227–2234.

30. Gorzałczany, M.B.; Rudziński, F. A multi-objective-genetic-optimization-based data-driven fuzzy classifier for technical applications. In Proceedings of the 2016 IEEE 25th International Symposium Industrial Electronics (ISIE), Santa Clara, CA, USA, 8–10 June 2016; pp. 78–83.

31. Xu, Y.; Dong, Z.Y.; Meng, K.; Zhang, R.; Wong, K.P. Real-time transient stability assessment model using extreme learning machine. *IET Gener. Transm. Distrib.* **2011**, *5*, 314–322. [CrossRef]

32. Menke, J.H.; Schäfer, F.; Braun, M. Performing a virtual field test of a new monitoring method for smart power grids. In Proceedings of the 2018 IEEE International Conference on Communications, Control, and Computing Technologies for Smart Grids, SmartGridComm 2018, Aalborg, Denmark, 29–31 October 2018; pp. 1–7.

33. Sogabe, T.; Malla, D.B.; Takayama, S.; Sakamoto, K.; Yamaguchi, K.; Singh, T.P.; Masaru, S. Smart grid optimization by deep reinforcement learning over discrete and continuous action space. In Proceedings of the 32nd Annual Conference of the Japanese Society for Artificial Intelligence (JSAI 2018), Kagoshima-shi, Japan, 5–8 June 2018; pp. 1–4.

34. Kosek, A. Contextual anomaly detection for cyber-physical security in smart grids based on an artificial neural network model. In Proceedings of the 2016 Joint Workshop on Cyber-Physical Security and Resilience in Smart Grids, Vienna, Austria, 12 April 2016; pp. 1–6.

35. Kaygusuz, C.; Babun, L.; Aksu, H.; Uluagac, S. Detection of compromised smart grid devices with machine learning and convolution techniques. In Proceedings of the 2018 IEEE International Conference Communications (ICC 2018), Kansas City, MO, USA, 20–24 May 2018; pp. 1–6.

36. Wang, B.; Fang, B.; Wang, Y.; Liu, H.; Liu, Y. Power system transient stability assessment based on big data and the core vector machine. *IEEE Trans. Smart Grid* **2016**, *7*, 2561–2570. [CrossRef]

37. Sajan, K.; Kumar, V.; Tyagi, B. Genetic algorithm based support vector machine for on-line voltage stability monitoring. *Electr. Power Energy Syst.* **2015**, *73*, 200–208. [CrossRef]

38. Malbasa, V.; Zheng, C.; Chen, P.C.; Popovic, T.; Kezunovic, M. Voltage stability prediction using active machine learning. *IEEE Trans. Smart Grid* **2017**, *8*, 3117–3124. [CrossRef]

39. Kofinas, P.; Dounis, A.; Vouros, G. Fuzzy Q-learning for multi-agent decentralized energy management in microgrids. *Appl. Energy* **2018**, *219*, 53–67. [CrossRef]

40. Japkowicz, N.; Shah, M. *Evaluating Learning Algorithms: A Classification Perspective*; Cambridge University Press: Cambridge, UK, 2011.

41. Gacto, M.J.; Alcala, R.; Herrera, F. Interpretability of linguistic fuzzy rule-based systems: an overview of interpretability measures. *Inf. Sci.* **2011**, *181*, 4340–4360. [CrossRef]

42. Okabe, T.; Jin, Y.; Sendhoff, B. A critical survey of performance indices for multi-objective optimisation. In Proceedings of the 2003 Congress on Evolutionary Computation, CEC '03, Canberra, Australia, 8–12 December 2003; Voume 2, pp. 878–885.

43. Panda, D.; Das, S. Regression analysis of grid stability under decentralized control. In Proceedings of the 2019 International Conference Engineering, Science, and Industrial Applications (ICESI), Tokyo, Japan, 22–24 August 2019; pp. 1–6.

44. Moldovan, D.; Salomie, I. Detection of sources of instability in smart grids using machine learning techniques. In Proceedings of the 2019 IEEE 15th International Conference Intelligent Computer Communication and Processing (ICCP), Cluj-Napoca, Romania, 5–7 September 2019; pp. 175–182.

45. Karim, J. Line Stability Analysis of the Decentral Smart Grid Control(DSGC). 2020. Available online: https://medium.com/analytics-vidhya/line-stability-analysis-of-the-decentral-smart-grid-control-d5ef7e94fe77 (accessed on 1 May 2020).

46. Balali, F.; Nouri, J.; Nasiri, A.; Zhao, T. *Data Intensive Industrial Asset Management: IoT-Based Algorithms and Implementation*; Springer: Cham, Switzerland, 2020.

47. Chen, H.; Wu, H.; Chan, S.; Lam, W. A Stochastic Quasi-Newton Method for Large-Scale Nonconvex Optimization with Applications. *IEEE Trans. Neural Netw. Learn. Syst.* **2019**, 1–15. [CrossRef]

48. Kingma, D.P.; Ba, J. Adam: A method for stochastic optimization. In Proceedings of the 3rd International Conference on Learning Representations, ICLR 2015, San Diego, CA, USA, 7–9 May 2015; Conference Track Proceedings. Bengio, Y., LeCun, Y., Eds.; pp. 1–15.

49. Arrieta, A.B.; Diaz-Rodriguez, N.; Del Ser, J.; Bennetot, A.; Tabik, S.; Barbado, A.; Garcia, S.; Gil-Lopez, S.; Molina, D.; Benjamins, R.; et al. Explainable artificial intelligence (XAI): Concepts, taxonomies, opportunities and challenges toward responsible AI. *Inf. Fusion* **2020**, *58*, 82–115. [CrossRef]

50. Samek, W.; Montavon, G.; Vedaldi, A.; Hansen, L.; Müller, K.R. (Eds.) *Explainable AI: Interpreting, Explaining and Visualizing Deep Learning*; Lecture Notes in Artificial Intelligence; Springer Nature Switzerland AG: Cham, Switzerland, 2019.

51. Monar, C. Interpretable Machine Learning. A Guide for Making Black Box Models Explainable. 2020, pp. 1–318. Available online: https://christophm.github.io/interpretable-ml-book/ (accessed on 1 May 2020).

52. Escalante, H.J.; Escalera, S.; Guyon, I.; Baró, X.; Güçlütürk, Y.; Güçlü, U.; Gerven, M. (Eds.) *Explainable and Interpretable Models in Computer Vision and Machine Learning*; The Springer Series on Challenges in Machine Learning; Springer Nature Switzerland AG: Cham, Switzerland, 2018.

Article

Time Series Clustering of Electricity Demand for Industrial Areas on Smart Grid

Heung-gu Son [1], Yunsun Kim [2] and Sahm Kim [2,*]

[1] Department of Short-term Demand Forecasting, Korea Power Exchange, Naju 58322, Korea; koreashg@kpx.or.kr
[2] Department of Applied Statistics, Chung-ang University, Seoul 06974, Korea; kys2125@cau.ac.kr
* Correspondence: sahm@cau.ac.kr; Tel.: +82-2-820-5225

Received: 30 March 2020; Accepted: 7 May 2020; Published: 9 May 2020

Abstract: This study forecasts electricity demand in a smart grid environment. We present a prediction method that uses a combination of forecasting values based on time-series clustering. The clustering of normalized periodogram-based distances and autocorrelation-based distances are proposed as the time-series clustering methods. Trigonometrical transformation, Box–Cox transformation, autoregressive moving average (ARMA) errors, trend and seasonal components (TBATS), double seasonal Holt–Winters (DSHW), fractional autoregressive integrated moving average (FARIMA), ARIMA with regression (Reg-ARIMA), and neural network nonlinear autoregressive (NN-AR) are used for demand forecasting based on clustering. The results show that the time-series clustering method performs better than the method using the total amount of electricity demand in terms of the mean absolute percentage error (MAPE).

Keywords: smart grid; DSHW; TBATS; NN-AR; time-series clustering

1. Introduction

In order to switch to a smart grid environment, real-time power demand data of each residence and industry is collected through the supply of AMI (advanced metering infrastructure). We aim to achieve efficient power use through an environment that can identify and control electricity consumption with AMI. After the expansion of smart-metering devices, it is necessary to forecast demand and supply accurately. The power system in a smart grid environment enhances the efficiency of power use and production using the exchange of information between the electricity supplier and the consumer through the smart meter [1].

In the Korea domestic electric power market, electrical power is produced while maintaining a reserve of 40,000 MW based on the maximum demand. However, after the "Second Energy Basic Plan" policy was promulgated in 2014 [2], the power management policy transformed from a power supply center policy into a power demand management center policy. Therefore, it is required to study the demand forecast based on the bottom-up method according to the individual companies and households, as well as power demand forecasts across the country. You should also consider environmental protection issues as well as energy efficiency issues.

In the power supply center scenario, efficiency was low owing to power being consumed even when it was not needed, and problems such as greenhouse gas emissions (which pose a serious threat to the environment) occurred owing to the burning of coal, oil, and gas [3]. A greenhouse gas (GHG) reduction target of 30%, as compared with the 2020 business-as-usual (BAU) levels, was fixed consequent to the G7 expansion summit in July 2008. Therefore, reducing emissions has become a mandatory requirement. It is possible to reduce greenhouse gas emissions through accurate electric power demand prediction, which helps avoid unnecessary electric power generation. A smart grid environment can result in improved accuracy of electric power demand prediction, by transitioning from the existing method of top-down prediction of the country's total electricity demand to a bottom-up approach; this improved accuracy will help reduce greenhouse gas emissions. In other words, by eliminating unnecessary power production through the introduction of high-efficiency power grids, which incorporate smart grids and smart meters, greenhouse gas emissions can be reduced.

Even though the importance of accurate prediction is emphasized in the smart grid environment, the development of the forecasting methodology is progressing somewhat slowly owing to the obstacle of private information disclosure. In addition, the name and the characteristics of the company are often not revealed even if the electricity usage is disclosed. In this study, despite the fact that electricity demand data for each company were collected, there was a problem that the characteristics of the company (e.g., corporate sector, contracted power volume, region, name, size of the buildings) could not be obtained.

Besides, even when collecting company-specific data, situations may arise where the classification criteria are ambiguous. We were only able to collect ID numbers and hourly power demand data for each company. Therefore, this study intends to present a forecasting method in a situation where only electricity demand data for each company are provided or in a condition where the classification of companies is obscure.

First, companies are clustered using the demand pattern of electricity demand. As the data of the same pattern fit for the same time-series model, companies belonging to the same cluster are aggregated into one cluster power data to make forecasts according to the appropriate model. Lastly, the total power demand is predicted by summing the results for each cluster.

In the past, before information and communication technologies (ICTs) and smart grids were developed, forecasting was based primarily on supply-side aggregated data in top-down formats at overarching governmental levels. Recently, owing to the development of computer technology, it has become possible to consider end-user demand through a bottom-up approach. The top-down approach was considered appropriate for the short-term load forecasting (STLF) method in the past. However, currently, the bottom-up approach is also applicable for STLF [4]. Thus, these technologies have expanded their roles by helping forecast peak load demand.

Forecasting methods are classified into statistical and non-statistical methods, depending on the underlying technique. Statistical methods generate mathematical equations from existing historical data to estimate model parameters and obtain predictions. These methods include autoregressive integrated moving average (ARIMA) models [5,6], regression seasonal autoregressive moving average generalized autoregressive conditional heteroscedasticity models (Reg-SARIMA-GARCH models) [7,8], exponential smoothing methods [9,10], time-series models for series exhibiting multiple complex seasonality (trigonometrical transformation, Box–Cox transformation, ARMA errors, trend and seasonal components (TBATS)) [10,11], regression models [12], support vector machine (SVM) models [13,14], fuzzy models [15,16], and Kalman filters [17].

In contrast, Artificial Intelligence (AI)-based techniques have high predictive power and are suitable for nonlinear data because of their nonlinear and nonparametric function characteristics. Many studies using neural network models such as convolutional neural network (CNN) [18,19], recurrent neural network (RNN) [20,21], and long short-term memory (LSTM) [22,23] have been published. Recent studies on the smart grid environment are briefly reviewed in the following paragraphs. Mohammadi et al. [24] proposed a hybrid model consisting of an improved Elman neural network (IENN) and novel shark smell optimization (NSSO) algorithm based on maximizing relevancy and minimizing redundancy to predict smart-meter data in Iran. Ghadimi et al. [25] studied a hybrid forecast model of artificial neural networks (ANNs), radial basis function neural networks (RBFNNs), and support vector machines (SVMs) to predict loads and prices in smart grids, based on Australia and new England data. They showed that the proposed model, using a dual-tree complex wavelet transform and a multi-stage forecast engine, was superior across the four seasons, as compared with other classic models. Kim et al. [26] provided advanced metering infrastructure (AMI) forecasting data for Korea using a long short-term memory (LSTM) network with a sequence of past profiles, temperature, and humidity.

Vrablecova et al. [27] used the support vector regression (SVR) method to predict smart grid data for 5000 households in Ireland. They compared various forecasting methods using an optimal window, and the results showed that the online SVR method was suitable for STLF in non-stationary data. Chou et al. [28] forecasted the energy consumption of air conditioners from a smart grid system using a hybrid model of ARIMA, metaheuristic firefly algorithm (MetaFA), and least-squares support vector regression (LSSVR) to integrate linear and nonlinear characteristics. The models were compared for various sizes of the sliding window as well as various input factors, and showed superiority over the basic models. Mohammad et al. [29] studied smart meters from 100 commercial buildings using the ARIMA model, exponential smoothing method, and seasonal and trend decomposition using loess (SLT) method. The buildings were compared based on industry characteristics, and the 5 min recorded data were aggregated with the 30 min data to ensure that a similar day approach could be followed. Muralitharan et al. [30] compared the neural network-based optimization approach for smart grid prediction. The results showed that the neural network based genetic algorithm (NNGA) was suitable for STLF. At the same time, neural network-based particle swarm optimization (NNPSO) was better for the long-term load forecast (LTLF). Kim et al. [8] compared multiple time-series methods (SARIMA, GARCH-ARIMA, and exponential smoothing methods) and AI-based (ANN) methods in a comprehensive approach for STLF over 1 h to 1 day ahead forecasting horizons. They showed that the optimal model was the ANN model with external variables of weather and holiday effects over the time horizons. Kundu et al. [31] worked on the uncertainty of parameters and measurements for hourly energy consumption forecasts. They analyzed the sensitivity of the optimization with commercial heating ventilation and air conditioning (HVAC) system data. Ahmad et al. [32] proposed a novel forecasting technique for a one-day ahead prediction in a smart grid environment with an intelligent modular approach. Besides, in the smart grid environment, prediction accuracy and calculation time, which is a trade-off relationship, are essential to consider. Amin et al. [33] compared a linear regression model, univariate seasonal ARIMA model, and the novel multivariate LSTM model for 114 residential apartment smart meters over two years. The performances varied from each model by seasons and forecasting horizons. Overall, the LSTM model was the most accurate; however, under the high variability seasons of the temperature, the simple regression model was better. Motlagh et al. [34] recommended a clustering method to support electricity smart grid forecasts of a large residential dataset. To overcome the unequal time-series of each residential smart meter, they suggested model-based clustering to compute parallel data for large samples. Moreover, the results of clustering were described as intra-cluster consistency and variability factors.

The authors reviewed numerous studies about the smart grid and forecasting methods, however, most of the studies dealt with aggregated data or independent small area data. In a smart grid environment, the government recommends the bottom-up method for figuring out the various energy-consuming patterns in the economic distribution system. Nevertheless, it is time-consuming to build up the independent forecasting models for each grid. Therefore, we would like to suggest clustering methods to forecast the AMI systems efficiently. There have been some papers proposing a clustering method in smart grid datasets [34], but the study was conducted in residential smart meters. The present study makes the following contributions. We confirmed the robustness of the bottom-up method using AMI data in industrial areas on the smart grid with time-series clustering methodologies. There are some differences in time-series patterns for residents and industries, and the total amount of energy consumptions are a more substantial portion in industry areas. Moreover, forecasting through time-series pattern analysis was used to reduce the time of computing and make the method more natural to use in practice.

The remainder of this paper is organized as follows. Section 2 introduces the methodology used in this study. Section 3 describes the data and variables and discusses the application of the data to the models, and Section 4 concludes the paper.

2. Methodology

In this study, we present bottom-up demand forecasting using time-series data for industrial sectors collected through AMI. Figure 1 describes a schematic format of the forecasting design. The AMI data used are electricity demand by the company per hour and use only the ID and power demand of the company. First, companies are clustered using demand patterns of electricity demand for each AMI data. For clustering by company, time-series cluster analysis, which is a cluster analysis method according to a similar time-series pattern, is used. The time-series clustering method considered is presented in Section 2.1. Next, we estimate the optimal model of the sum of the data in the cluster. The considered time-series prediction model is presented in Section 2.2. Finally, the total amount of power generation of the entire company is estimated by summing up the optimal model prediction values for each cluster.

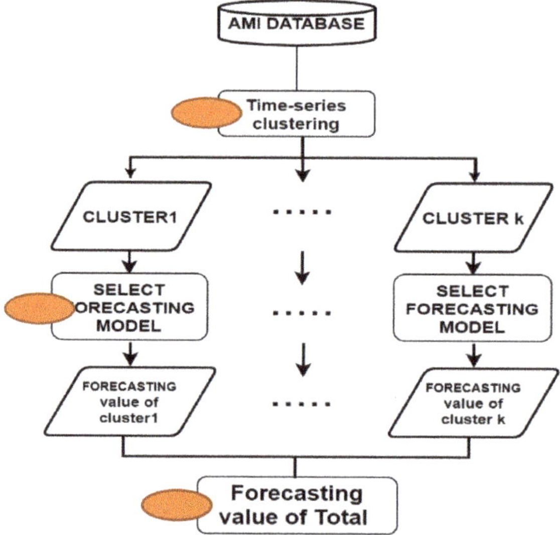

Figure 1. Methodology for clustering of electricity demands for industrial areas. AMI, advanced metering infrastructure.

2.1. The Time-Series Clustering

2.1.1. Autocorrelation-Based Distances

Galeano et al. [35] suggested a clustering technique for autocorrelation function (ACF) in the time-series data. $\hat{\rho}_{X_T} = \left(\hat{\rho}_{1,X_T}, \cdots, \hat{\rho}_{L,\,X_T}\right)'$ and $\hat{\rho}_{Y_T} = \left(\hat{\rho}_{1,Y_T}, \cdots, \hat{\rho}_{L,Y_T}\right)'$ are estimated autocorrelation vectors of X_T and Y_T, respectively, then $\hat{\rho}_{1,X_T} \approx 0$ and $\hat{\rho}_{1,Y_T} \approx 0$ when $i > L$. The ACF measures are expressed as below.

$$d_{\text{ACF}}(X_T, Y_T) = \sqrt{\left(\hat{\rho}_{X_T} - \hat{\rho}_{Y_T}\right)' \Omega \left(\hat{\rho}_{X_T} - \hat{\rho}_{Y_T}\right)} \tag{1}$$

where Ω is a weight matrix, and the equal weights are assumed by giving the initial Ω as I. In this case, d_{ACF} becomes a Euclidiean distance measure between estimated ACF, as below.

$$d_{\text{ACF}}(X_T, Y_T) = \sqrt{\sum_{i=1}^{L}\left(\hat{\rho}_{1,X_T} - \hat{\rho}_{1,Y_T}\right)^2} \tag{2}$$

The measure can be expressed as below as the considering geometric weights without time lag in ACF.

$$d_{\text{ACF G}}(X_T, Y_T) = \sqrt{\sum_{i=1}^{L} p(1-p)^i \left(\hat{\rho}_{1,X_T} - \hat{\rho}_{1,Y_T}\right)^2}, \quad \text{with } 0 < p < 1 \tag{3}$$

2.1.2. Normalized Periodogram-Based Distance

The periodogram distance proposed by Caiado et al. [36] is a metric to recognize the keyframes using a short boundary estimated on a sliding sub-window basis. If its correlation structure is more appropriate than the process scale, then it is better to use the normalized periodogram by using the Euclidean distance. The normalized periodogram measures are expressed as below.

$$d_{\text{NP}}(X_T, Y_T) = \frac{1}{n} \sqrt{\sum_{k=1}^{N} \left(NI_{X_T}(\lambda_k) - NI_{Y_T}(\lambda_k)\right)^2} \tag{4}$$

where $NI_{X_T}(\lambda_k) = T^{-1} \left| \sum_{t=1}^{T} X_t e^{-i\lambda_k t} \right|^2$ and $NI_{Y_T}(\lambda_k) = T^{-1} \left| \sum_{t=1}^{T} Y_t e^{-i\lambda_k t} \right|^2$ are the periodograms of X_t and Y_t, respectively. $\lambda_k = \frac{2\pi k}{T}$, $k = 1, \ldots, n$.

2.2. Forecasting Models

The double seasonal Holt–Winters (DSHW), trigonometric transform, Box–Cox transform, ARMA errors, trend and seasonal components (TBATS), fractional autoregressive integrated moving average (FARIMA), ARIMA with regression (Reg-ARIMA), and neural network nonlinear autoregressive (NN-AR) models are presented in this section.

2.2.1. The Double Seasonal Holt–Winters (DSHW) Method

The extension version for the double seasonal Holt–Winters method helped address multiple seasonal cycles, and can be written as below [37].

$$L_t = \alpha(y_t - S_{t-s_1} - D_{t-s_2}) + (1-\alpha)(L_{t-1} + T_{t-1}) \tag{5}$$

$$T_t = \beta(L_t - L_{t-1}) + (1-\beta) T_{t-1} \tag{6}$$

$$S_t = \gamma(y_t - L_t - D_{t-s_2}) + (1-\gamma)S_{t-s_1} \tag{7}$$

$$D_t = \delta(y_t - L_t - S_{t-s_1}) + (1-\delta)D_{t-s_2} \tag{8}$$

$$F_{t+h} = L_t + T_t \times h + S_{t+h-s_1} + D_{t+h-s_2}, \tag{9}$$

where y_t represents the actual data; S_t, D_t represent the seasonal component over time t ($t = 1, 2, \cdots, T$); and s_1, s_2 are the double seasonal cycles. The components L_t and T_t describe the level and trend of the series at time t, respectively. The coefficients α, β, γ are parameters for smoothing. F_{t+h} is the forecasting value of h ahead of time t. The initial points are calculated as follows.

$$L_{s_1} = \frac{1}{s_1} \sum_{t=1}^{s_1} y_t, \quad L_{s_2} = \frac{1}{s_2} \sum_{t=1}^{s_2} y_t \tag{10}$$

$$T_{s_1} = \frac{1}{s_1^2} \left(\sum_{t=s_1+1}^{2s_1} y_t - \sum_{t=1}^{s_1} y_t \right), \quad T_{s_2} = \frac{1}{s_2^2} \left(\sum_{t=s_2+1}^{2s_2} y_t - \sum_{t=1}^{s_2} y_t \right) \tag{11}$$

$$S_1 = y_1 - L_{s_1}, \cdots, S_{s_1} = y_{s_1} - L_{s_1} \tag{12}$$

$$D_1 = y_1 - L_{s_2}, \cdots, S_{s_2} = y_{s_2} - L_{s_2} \tag{13}$$

2.2.2. Trigonometric Transform, Box–Cox Transform, ARMA Errors, Trend and Seasonal Components (TBATS) Model

TBATS is an acronym for trigonometrical transformation, Box–Cox transformation, ARMA errors, trend and seasonal components. To overcome the problems of a wider seasonality and correlated errors in the exponential smoothing method, modified state-space models were introduced by De Livera et al. [38]. It is restricted to linear homoscedasticity, but the Box–Cox transformation can handle some types of non-linearity. This class of model is named BATS (Box–Cox transformation, ARMA errors, trend and seasonal components) and is defined as follows.

$$y_t^{(\omega)} = \left\{ \begin{array}{ll} \frac{y_t^{\omega} - 1}{\omega}, & \omega \neq 0 \\ log(y_t), & \omega = 0 \end{array} \right\} \tag{14}$$

$$y_t^{(\omega)} = l_{t-1} + \phi b_{t-1} + \sum_{i=1}^{T} s_{t-m_1}^{(i)} + d_t \tag{15}$$

$$l_t = l_{t-1} + \phi b_{t-1} + \alpha d_t \tag{16}$$

$$b_t = (1 - \phi)b + \phi b_{t-1} + \beta d_t \tag{17}$$

$$S_t^{(i)} = S_{t-m_i}^{(i)} + \gamma_i d_t \tag{18}$$

$$d_t = \sum_{i=1}^{p} \varphi_i d_{t-i} + \sum_{i=1}^{q} \theta_i \varepsilon_{t-i} + \varepsilon_t \tag{19}$$

where $y_t^{(\omega)}$ is the Box–Cox transformed data for parameter ω at time t, l_t depicts the local level, b is the long-term trend, and b_t is the short-term trend within the period of time. Rather than converging on zero, the value of b_t finally meets on b. ϕ is a damping parameter for the trend. d_t is a series of ARMA models with orders (p, q) and ε_t is the white noise process with a mean of zero and a constant variance of σ^2. m_i is the ith seasonal cycle. α, β, and γ_i are the smoothing parameters for $i = 1, \cdots, T$.

The trigonometric seasonal approach incorporated into the model leads to a reduction in the estimation time (which increases with the number of parameters). This approach further accommodates non-integer seasonality. The final arguments for TBATS model $(\omega, \phi, p, q, \{m_1, k_1\}, \{m_2, k_2\}, \cdots, \{m_T, k_T\})$ are explained with some additional equations, as below.

$$S_t^{(i)} = \sum_{j=1}^{k_i} S_{j,t}^{(i)} \tag{20}$$

$$S_{j,t}^{(i)} = S_{j,t-1}^{(i)} cos\lambda_j^{(i)} + S_{j,t-1}^{*(i)} sin\lambda_j^{(i)} + \gamma_1^{(i)} d_t \tag{21}$$

$$S_{j,t}^{*(i)} = -S_{j,t-1}^{(i)} sin\lambda_j^{(i)} + S_{j,t-1}^{*(i)} cos\lambda_j^{(i)} + \gamma_2^{(i)} d_t, \tag{22}$$

where k_i is the number of harmonics for $S_t^{(i)}$, which is a seasonal component. $\gamma_1^{(i)}$ and $\gamma_2^{(i)}$ are the smoothing parameters and $\lambda_j^{(i)} = \frac{2\pi j}{m_i}$. $S_{j,t}^{(i)}$ is the stochastic level of the ith seasonal component by $S_{j,t}^{(i)}$, and $S_{j,t}^{*(i)}$ is the stochastic growth of the ith seasonal component.

2.2.3. The Fractional Autoregressive Integrated Moving Average (FARIMA) Model

The FARIMA model is the common generalization of regular ARIMA processes when the degree of differencing d can take nonintegral values [39]. When the series $\{y_t | t = 1, 2, \cdots, T\}$ follows ARIMA (p, d, q), the time series takes the following form:

$$\phi_p(l)(1-l)^d y_t = \theta_q(l)\varepsilon_t, \tag{23}$$

where y_t denotes the actual data observed at time t ($t = 1, 2, \cdots, T$), and ε_t describes the random errors assuming white noise on t with a mean of zero and a constant variance of σ^2. p and q are integers and orders in the model. $\phi_p(l) = 1 - \phi_1 l - \cdots - \phi_p l^p$, where p represents the degree of the autoregressive polynomial. $\theta_q(l) = 1 - \theta_1 l - \cdots - \theta_q l^q$, where q is the degree of the moving average polynomial. $(1-l)^d = \sum_{j=1}^{\infty} \binom{d}{j}(-1)^j L^j$ with $\binom{d}{j}(-1) = \frac{\Gamma(-d+j)}{\Gamma(-d)\Gamma(j+1)}$, and $d \in (-0.5, \ 0.5)$.

2.2.4. Reg-ARIMA

The Reg-ARIMA model is proposed by Bell and Hilmer [40], it is a compound word of Regression and ARIMA. We consider hourly temperature data as a regressor, and double seasonality was fitted to explain the daily and weekly cycles. When the series $\{y_t | t = 1, 2, \cdots, T\}$ follows ARIMA $(p, d, q)(P_1, D_1, Q_1)_{s_1}(P_2, D_2, Q_2)_{s_2}$, the time series takes the following form:

$$\phi_p(l)\Phi_{P_1}(l^{s_1})\Pi_{P_2}(l^{s_2})(1-l)^d(1-l^{s_1})^{D_1}(1-l^{s_2})^{D_2}\left(y_t - \sum_{j=1}^{r} \beta_j x_{jt}\right) \tag{24}$$
$$= \theta_q(l)\Theta_{Q_1}(l^{s_1})\Psi_{Q_2}(l^{s_2})\varepsilon_t,$$

where β_j is a coefficient for the j-th regressor, y_t denotes the actual data observed at time t ($t = 1, 2, \cdots, T$), and ε_t describes the random errors assuming white noise on t with a mean of zero and a constant variance of σ^2. p and q are integers and orders in the model. $\phi_p(l) = 1 - \phi_1 l - \cdots - \phi_p l^p$, where p represents the degree of the autoregressive polynomial. $\theta_q(l) = 1 - \theta_1 l - \cdots - \theta_q l^q$, where q is the degree of the moving average polynomial. Moreover, for the first seasonal operators, $\Phi_{P_1}(l^{s_1}) = 1 - \Phi_1 l^{s_1} - \cdots - \Phi_{P_1} l^{Ps_1}$, and $\Theta_{Q_1}(l^{s_1}) = 1 - \Theta_1 l^{s_1} - \cdots - \Theta_{Q_1} l^{Qs_1}$, where P_1 and Q_1 are the degree of the first-seasonal autoregressive polynomial and moving average polynomial, respectively. For the second seasonal operators, $\Pi_{P_2}(l^{s_2}) = 1 - \Pi_1 l^{s_2} - \cdots - \Pi_{P_2} l^{Ps_2}$, and $\Psi_{Q_2}(l^{s_2}) = 1 - \Psi_1 l^{s_2} - \cdots - \Psi_{Q_2} l^{Qs_2}$, where P_2 and Q_2 are the degree of the second-seasonal autoregressive polynomial and moving average polynomial, respectively. $(1-l)^d$, $(1-l^{s_1})^{D_1}$, and $(1-l^{s_2})^{D_2}$ are the non-seasonal, first, and second seasonal difference operators of order d and D, respectively. s_1 and s_2 represent a seasonal cycle.

2.2.5. Neural Network Nonlinear Autoregressive (NN-AR)

Artificial neural network (ANN) models are designed similar to the neurons of a human brain. There are substantial complex forms of connected neurons in human brains. The network cells carry a specific signal from the body along an axon, transferring the signals to other neurons. The connections among axons are processed by a synapse. Some neurons are structured at birth, some grow and mature, and the rest die when considered to be non-useful. Likewise, the neural network model having a single-input neuron is defined below [41].

$$a = f(wp + b) \tag{25}$$

where p is an input, w is the weight of p, and b is a bias. f is a transfer function used to obtain an output of a, and it can be either linear or non-linear. If the neuron structure consists of R numbers of inputs as $p_1, p_2, \cdots p_R$, the expression of summarized inputs can be expressed as follows:

$$n = w_{1,1}p_1 + w_{1,2}p_2 + \cdots + w_{1,R}p_R + b \tag{26}$$

where $w_{i,j}$ is the connection weight of the ith neuron from the jth neuron. n can also be expressed as a matrix form of inputs and weights: $n = \mathbf{Wp} + b$. If there are S neurons in a layer, the model can be expressed as follows: $\mathbf{a} = \mathbf{f}(\mathbf{Wp} + \mathbf{b})$, where $\mathbf{b}$ is a bias vector, $\mathbf{a}$ is an output vector, and $\mathbf{p}$ is an input vector with the weight matrix, $\mathbf{W}$. The matrix, $\mathbf{W}$, can be written as follows:

$$\mathbf{W} = \begin{bmatrix} w_{1,1} & \cdots & w_{1,R} \\ \vdots & \ddots & \vdots \\ w_{S,1} & \cdots & w_{S,R} \end{bmatrix} \tag{27}$$

If we expand the single layer to multiple layers, the output of the first layer can become an input to the next layer. For example, the output from the three layers can be explained as follows:

$$\mathbf{a}^3 = \mathbf{f}^3\left(\mathbf{W}^3\mathbf{f}^2\left(\mathbf{W}^2\mathbf{f}^1\left(\mathbf{W}^1\mathbf{p} + \mathbf{b}^1\right) + \mathbf{b}^2\right) + \mathbf{b}^3\right) \tag{28}$$

In this case, the third layer is considered as the output layer, and the first and second layers are hidden layers. The model is advantageous, because it can explain the complex relationships of the neurons with nonlinear functions. NN-AR models when additional exogenous variables are used for the ANN-based model.

3. Application of the Models

The auto-meter-reading (AMR) data obtained from Korea Electric Power Corporation (Naju, Korea) comprise data for 114 commercial buildings [42]. The data we provided were already refined; therefore, there was no extra step for data preprocessing. They were collected at 1 h intervals during the period from 12 February to 28 April 2014. A period of 10 weeks (10 February–20 April) was used for training, and the remaining one week (21 April–27 April) was used for testing. The datasets from the four weeks (24 March–20 April) prior to the test set were used for cluster analysis. Figure 2 shows an average time series profile of 114 buildings. The demand shows daily seasonal patterns, and there is a pattern of increasing demand from February to March.

We fitted the data with DSHW, TBATS, FARIMA, Reg-ARIMA, ANN, and NN-AR models; regarding the clustering, we chose the model showing the lowest mean absolute percentage error (MAPE) in each cluster. Then, the forecasting values from the aggregated and clustered data were compared. The model performances were evaluated using the mean absolute percentage error (MAPE). These evaluation methods are widely used to evaluate model performance, especially for STLF.

MAPE is defined as follows:

$$\text{MAPE} = \frac{100}{n} \sum_{t=1}^{n} \left| \frac{y_t - \hat{y}_t}{y^t} \right| \tag{29}$$

where y_t is the actual value and $\hat{y}_t$ is the forecasted demand at time t.

The number of clusters was chosen based on the silhouette metric. Figure 3 shows the time-series profiles of each of the 10 clusters measured. It demonstrates different patterns that have their own periodicity in the clusters. In this study, the double seasonality of daily and weekly patterns is considered. The optimal number of parameters at each k step in FARIMA models and Reg-ARIMA models was identified according to the corrected Akaike information criterion (AICc). Table 1 presents the FARIMA models in the training set of the aggregated data. The optimal parameter of d was selected as 0.1206 from the range $(-0.5, 0.5)$, and then the number of p and q was selected as 5 and 5, respectively. Table 2 shows the identification of the best Reg-ARIMA models. Tables 3 and 4 represent the hyperparameter tuning in ANN and NN-AR models, respectively. As we used nnetar function in the forecast package of R for the hyperparameter optimization in ANN and NN-AR models, we were able to search the number of nodes in the hidden layer, weight decay, iteration times, and network numbers. The optimized hyperparameters are chosen, indicating the minimum sum of squared errors.

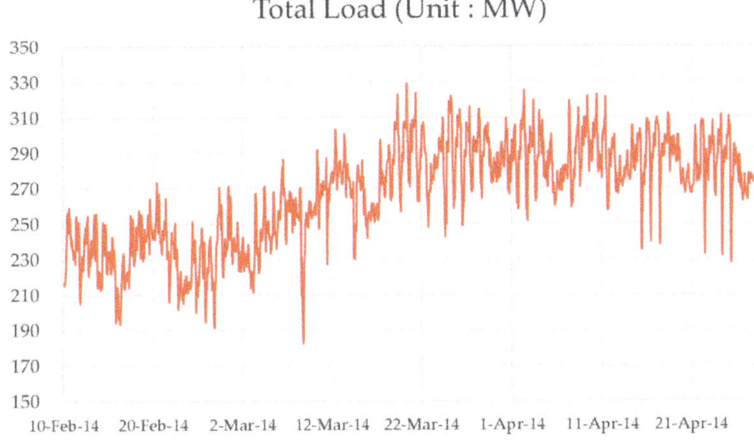

Figure 2. Average demand plot from the 114 buildings.

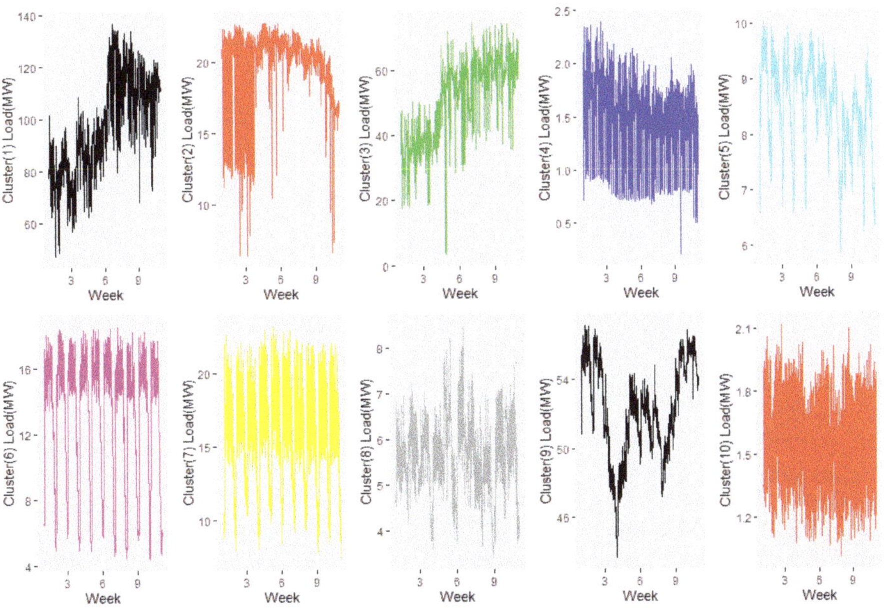

Figure 3. Demand plot for 10 cluster groups by autocorrelation-based distances.

Table 1. Identification of the best fractional autoregressive integrated moving average (FARIMA) model with $\hat{d} = 0.1206$. AICc, corrected Akaike information criterion.

(*p*,*q*)	AICc
5, 5	34,868.8
6, 4	34,873.3
7, 4	34,874.2
5, 4	34,875.4
6, 3	34,879.9

Table 2. Identification of the best regression ARIMA (Reg-ARIMA) model.

(*p*,*d*,*q*) (*P*,*D*,*Q*)$_{s=128}$	AICc
(2, 1, 3) (0, 1, 0)	28,177.9
(1, 1, 2) (0, 1, 0)	28,189.8
(2, 1, 2) (0, 1, 0)	28,190.6
(1, 1, 3) (0, 1, 0)	28,191.7
(3, 1, 2) (0, 1, 0)	28,192.1

Table 3. Hyperparameter optimization of the best artificial neural network (ANN) model.

Parameter	Search Space	Selected Value
Node numbers	(10, 15, 20, 25, 30)	15
Weight decay	(0.0, 0.05, 0.1)	0.0
Iteration times	(50, 100, 150, 200)	200
Network numbers	(15, 20, 25, 30)	20
Sum of squared errors		8,724,653

Table 4. Hyperparameter optimization of the best neural network nonlinear autoregressive (NN-AR) model.

Parameter	Search Space	Selected Value
Node numbers	(10, 15, 20, 25, 30)	15
Weight decay	(0.0, 0.05, 0.1)	0.0
Iteration times	(50, 100, 150, 200)	200
Network numbers	(15, 20, 25, 30)	15
Sum of squared errors		8,250,619

Tables 5–8 represent the estimated parameters and the results for assumptions in the training set. Figure 4 describes a dendrogram from the autocorrelation-based distances clustering. Table 9 presents the results of the MAPE in the test set and the number observations assigned in each cluster. It shows that the TBATS model is superior to other models in clusters 5 and 9 and the DSHW model elicits high accuracy in cluster 6. For clusters 1, 2, 3, 4, 7, 8, and 10, NN-AR was the most suitable model. Therefore, each cluster yields forecasting values from the different models. Further, we fit the NN-AR model for the aggregated data because it was superior to the other models in the dataset. Consequently, a week long forecasting was conducted for each cluster and total usage. In Table 10, the results indicate the MAPE of the aggregated forecasting value and the total usage data for the forecasted amounts for each cluster through the cluster analysis method. The MAPE for the forecasting of total data by the NN-AR model is 3.86%. The MAPE for the forecasting of aggregated data by clustering of autocorrelation-based distances is 3.32%, which is the summation forecasting result based on cluster-specific forecasting; the result based on clustering of normalized periodogram-based distance is 3.94%; the forecasting through clustering of autocorrelation-based distances showed a more accurate overall forecasting.

Table 5. Parameter estimations of double seasonal Holt–Winters (DSHW).

Parameter	Estimate
Level (α)	0.3022
Trend (β)	0.0003
Seasonal 1 (γ)	0.1630
Seasonal 2 (δ)	0.2467
ϕ	0.6988

Table 6. Parameter estimations of the TBATS model.

Parameter	Estimate
Level (α)	0.0245
ϕ_1	1.0765
ϕ_2	−0.2735
θ_1	0.0158
$\gamma_{1(24)}$	0.0004
$\gamma_{1(168)}$	−0.0006
$\gamma_{2(24)}$	0.0001
$\gamma_{2(168)}$	0.0002

Table 7. Parameter estimations of the FARIMA model.

Parameter	Estimate
ϕ_1	0.7254
ϕ_2	0.0243
ϕ_3	0.4461
ϕ_4	0.2097
ϕ_5	−0.4065
θ_1	−0.2022
θ_2	0.0014
θ_3	0.3324
θ_4	0.7423
θ_5	0.0588

Table 8. Parameter estimations of the Reg-ARIMA model.

Parameter	Estimate
ϕ_1	−0.1978
ϕ_2	0.7702
θ_1	0.3379
θ_2	−0.9602
θ_3	−0.3277
Temperature	−562.3

Table 9. Mean absolute percentage error (MAPE) of training dataset by autocorrelation-based distances clustering approach.

MAPE (%)	C1	C2	C3	C4	C5	C6	C7	C8	C9	C10
DSHW	3.5	4.3	7.5	4.9	1.5	1.7	3.4	6.6	0.5	9.9
TBATS	3.2	5.7	5.2	5.5	1.4	2.4	4.5	3.2	0.4	9.4
FARIMA	3.4	5.6	5.2	9.4	1.7	3.1	8.6	4.1	0.5	11.0
Reg-ARIMA	3.5	5.6	5.4	5.3	1.7	1.9	3.7	4.0	0.5	11.5
ANN	2.1	1.8	2.7	3.1	1.6	2.4	2.3	2.1	0.4	4.8
NN-AR	2.0	1.7	2.5	3.0	1.6	2.4	2.3	2.1	0.4	4.2
Obs.	29	5	13	2	7	14	24	9	6	5

Table 10. Result of forecasting accuracy by MAPE.

Method	Total (NN-AR)	$d_{\text{ACF G}}$	d_{NP}
MAPE (%)	3.86	3.32	3.94

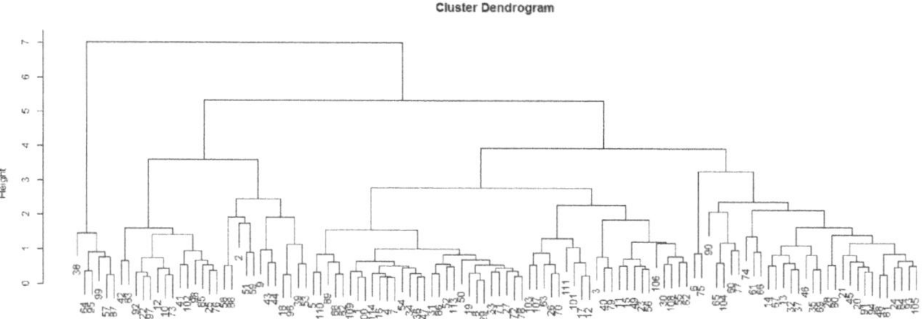

Figure 4. Dendrogram for autocorrelation-based clustering.

4. Conclusions

It is well known that forecasts based on the bottom-up method are more accurate than forecasts based on the top-down method if companies' power demand pattern is constant. However, in the actual operating establishment, as each company's prediction can cause problems such as excessive computing time, it is necessary to determine the most appropriate method.

This study presents a bottom-up power prediction method through the results of a cluster analysis of electricity demand for each company in a smart grid environment. To this end, data from 114 companies recorded using auto-metering-reading (AMR) devices were classified into 10 clusters using time series cluster analysis. Owing to the strong time-series characteristics of power demand, this study uses clustering of autocorrelation-based distances and normalized periodogram distance. The power demand classified through time-series cluster analysis revealed different patterns according to the characteristics of the companies. As the demand pattern for each cluster was different, the optimal forecasting model for each cluster was selected using the NN-AR and TBATS models. Finally, the total power demand was calculated based on the aggregated forecasting results for each cluster.

The forecasting result using the time-series cluster (autocorrelation-based distances) analysis of the bottom-up method was superior to that of the top-down method by about 0.5%.

This study proposed a solution to the problem of low predictive accuracy and computing speed owing to big data management in the process of transitioning to the smart grid method. If the power consumption of all companies is forecasting, accuracy will increase, but the program run time will be longer. In addition, if all companies and households with electricity demand are considered, the electricity demand forecasting will take longer. Accordingly, in real-time operations, it is proposed to improve the accuracy of prediction through time-series cluster analysis and to reduce the program run time by predicting each cluster analysis result. This method is proposed as a proactive response to issues related to the smart grid method with regard to national power management.

The results of this study may be relevant only to the demand data of generic companies. Therefore, any future analysis based on cluster characteristics may consider specificities. Future studies should examine the patterns of demand based on company characteristics by analyzing corporate information. Further, demand patterns should be examined according to the characteristics of the company's industry, and researchers should determine the optimal model for each cluster characteristic. Finally, it is necessary to determine a system automation method that can be used in an actual power operation establishment.

Author Contributions: Conceptualization, H.-g.S. and S.K.; Software, Y.K.; Formal Analysis, H.-g.S.; Formal Analysis, H.-g.S. and S.K.; Investigation, Y.K.; Project administration, S.K.; Supervision, S.K.; Writing—original draft, H.-g.S.; Writing—review & editing, Y.K.; Visualization, H.-g.S. All authors have read and agreed to the published version of the manuscript.

Funding: This research was funded by the National Research Foundation of Korea grant number 2016R1D1A1B01014954.

Acknowledgments: The authors are appreciative of the worthy commentaries and advice from the respected reviewers. Their helpful remarks have improved the power and importance of our paper.

Conflicts of Interest: The authors declare no conflict of interest.

References

1. Huang, A.Q. Power semiconductor devices for smart grid and renewable energy systems. In *Power Electronics in Renewable Energy Systems and Smart Grid*; Wiley: Hoboken, NJ, USA, 2019; pp. 85–152.
2. Korea Power Exchange. Available online: http://www.kpx.or.kr (accessed on 9 May 2020).
3. Renn, O.; Marshall, J.P. Coal, nuclear and renewable energy policies in Germany: From the 1950s to the "Energiewende". *Energy Policy* **2016**, *99*, 224–232. [CrossRef]
4. Reyna, J.; Chester, M.V. Energy efficiency to reduce residential electricity and natural gas use under climate change. *Nat. Commun.* **2017**, *8*, 14916. [CrossRef]
5. Alberg, D.; Last, M. Short-term load forecasting in smart meters with sliding window-based ARIMA algorithms. *Vietnam. J. Comput. Sci.* **2018**, *5*, 241–249. [CrossRef]
6. Nie, H.; Liu, G.; Liu, X.; Wang, Y. Hybrid of ARIMA and SVMs for Short-Term Load Forecasting. *Energy Procedia* **2012**, *16*, 1455–1460. [CrossRef]
7. Sigauke, C.; Chikobvu, D. Prediction of daily peak electricity demand in South Africa using volatility forecasting models. *Energy Econ.* **2011**, *33*, 882–888. [CrossRef]
8. Kim, Y.; Son, H.-G.; Kim, S. Short term electricity load forecasting for institutional buildings. *Energy Rep.* **2019**, *5*, 1270–1280. [CrossRef]
9. Chan, K.Y.; Dillon, T.S.; Singh, J.; Chang, E. Neural-Network-Based Models for Short-Term Traffic Flow Forecasting Using a Hybrid Exponential Smoothing and Levenberg–Marquardt Algorithm. *IEEE Trans. Intell. Transp. Syst.* **2011**, *13*, 644–654. [CrossRef]
10. Sohn, H.-G.; Jung, S.-W.; Kim, S. A study on electricity demand forecasting based on time series clustering in smart grid. *J. Appl. Stat.* **2016**, *29*, 193–203. [CrossRef]
11. Brożyna, J.; Mentel, G.; Szetela, B.; Strielkowski, W. Multi-Seasonality in the TBATS Model Using Demand for Electric Energy as a Case Study. *Econ. Comput. Econ. Cybern. Stud. Res.* **2018**, *52*, 229–246. [CrossRef]
12. Dudek, G. Pattern-based local linear regression models for short-term load forecasting. *Electr. Power Syst. Res.* **2016**, *130*, 139–147. [CrossRef]
13. Cao, G.; Wu, L.-J. Support vector regression with fruit fly optimization algorithm for seasonal electricity consumption forecasting. *Energy* **2016**, *115*, 734–745. [CrossRef]
14. Barman, M.; Choudhury, N.D.; Sutradhar, S. A regional hybrid GOA-SVM model based on similar day approach for short-term load forecasting in Assam, India. *Energy* **2018**, *145*, 710–720. [CrossRef]
15. Bin Song, K.; Baek, Y.-S.; Hong, D.H.; Jang, G. Short-Term Load Forecasting for the Holidays Using Fuzzy Linear Regression Method. *IEEE Trans. Power Syst.* **2005**, *20*, 96–101. [CrossRef]
16. Sadaei, H.J.; Guimarães, F.G.; Da Silva, C.J.; Lee, M.H.; Eslami, T. Short-term load forecasting method based on fuzzy time series, seasonality and long memory process. *Int. J. Approx. Reason.* **2017**, *83*, 196–217. [CrossRef]
17. Zheng, Z.; Chen, H.; Luo, X. A Kalman filter-based bottom-up approach for household short-term load forecast. *Appl. Energy* **2019**, *250*, 882–894. [CrossRef]
18. Dong, X.; Qian, L.; Huang, L. Short-term load forecasting in smart grid: A combined CNN and K-means clustering approach. In Proceedings of the 2017 IEEE International Conference on Big Data and Smart Computing (BigComp), Jeju Island, Korea, 13–16 February 2017.
19. Khan, S.; Javaid, N.; Chand, A.; Khan, A.B.M.; Rashid, F.; Afridi, I.U. Electricity load forecasting for each day of week using deep CNN. In *Proceedings of the Workshops of the International Conference on Advanced Information Networking and Applications*; Springer: Berlin/Heidelberg, Germany, 2019. [CrossRef]
20. Shi, H.; Xu, M.; Li, R. Deep Learning for Household Load Forecasting—A Novel Pooling Deep RNN. *IEEE Trans. Smart Grid* **2017**, *9*, 5271–5280. [CrossRef]
21. Yang, L.; Yang, H. Analysis of Different Neural Networks and a New Architecture for Short-Term Load Forecasting. *Energies* **2019**, *12*, 1433. [CrossRef]
22. Zheng, H.; Yuan, J.; Chen, L. Short-Term Load Forecasting Using EMD-LSTM Neural Networks with a Xgboost Algorithm for Feature Importance Evaluation. *Energies* **2017**, *10*, 1168. [CrossRef]

23. Bouktif, S.; Fiaz, A.; Ouni, A.; Serhani, M.A. Optimal Deep Learning LSTM Model for Electric Load Forecasting using Feature Selection and Genetic Algorithm: Comparison with Machine Learning Approaches. *Energies* **2018**, *11*, 1636. [CrossRef]

24. Mohammadi, M.; Talebpour, F.; Safaee, E.; Ghadimi, N.; Abedinia, O. Small-Scale Building Load Forecast based on Hybrid Forecast Engine. *Neural Process. Lett.* **2017**, *48*, 329–351. [CrossRef]

25. Ghadimi, N.; Akbarimajd, A.; Shayeghi, H.; Abedinia, O. A new prediction model based on multi-block forecast engine in smart grid. *J. Ambient. Intell. Humaniz. Comput.* **2017**, *9*, 1873–1888. [CrossRef]

26. Kim, N.; Kim, M.; Choi, J.K. Lstm based short-term electricity consumption forecast with daily load profile sequences. In Proceedings of the 2018 IEEE 7th Global Conference on Consumer Electronics (GCCE), Nara, Japan, 9–12 October 2018.

27. Vrablecová, P.; Ezzeddine, A.B.; Rozinajová, V.; Šárik, S.; Sangaiah, A.K. Smart grid load forecasting using online support vector regression. *Comput. Electr. Eng.* **2018**, *65*, 102–117. [CrossRef]

28. Chou, J.-S.; Hsu, S.-C.; Ngo, N.-T.; Lin, C.-W.; Tsui, C.-C. Hybrid Machine Learning System to Forecast Electricity Consumption of Smart Grid-Based Air Conditioners. *IEEE Syst. J.* **2019**, *13*, 3120–3128. [CrossRef]

29. Mohammad, R. AMI Smart Meter Big Data Analytics for Time Series of Electricity Consumption. In Proceedings of the 2018 17th IEEE International Conference on Trust, Security and Privacy in Computing And Communications/12th IEEE International Conference On Big Data Science And Engineering (TrustCom/BigDataSE), New York, NY, USA, 1–3 August 2018.

30. Muralitharan, K.; Sakthivel, R.; Vishnuvarthan, R. Neural network based optimization approach for energy demand prediction in smart grid. *Neurocomputing* **2018**, *273*, 199–208. [CrossRef]

31. Kundu, S.; Ramachandran, T.; Chen, Y.; Vrabie, D. Optimal Energy Consumption Forecast for Grid Responsive Buildings: A Sensitivity Analysis. In Proceedings of the 2018 IEEE Conference on Control Technology and Applications (CCTA), Copenhagen, Denmark, 21–24 August 2018.

32. Ahmad, A.; Javaid, N.; Mateen, A.; Awais, M.; Khan, Z.A. Short-Term Load Forecasting in Smart Grids: An Intelligent Modular Approach. *Energies* **2019**, *12*, 164. [CrossRef]

33. Amin, P.; Cherkasova, L.; Aitken, R.; Kache, V. Automating energy demand modeling and forecasting using smart meter data. In Proceedings of the 2019 IEEE International Congress on Internet of Things (ICIOT), Milan, Italy, 8–13 July 2019.

34. Motlagh, O.; Berry, A.; O'Neil, L. Clustering of residential electricity customers using load time series. *Appl. Energy* **2019**, *237*, 11–24. [CrossRef]

35. Galeano, P.; Peña, D. *Multivariate Analysis in Vector Time Series*; Universidad Carlos III de Madrid: Madrid, Spain, 2001.

36. Caiado, J.; Crato, N.; Peña, D. A periodogram-based metric for time series classification. *Comput. Stat. Data Anal.* **2006**, *50*, 2668–2684. [CrossRef]

37. Taylor, J.W. Triple seasonal methods for short-term electricity demand forecasting. *Eur. J. Oper. Res.* **2010**, *204*, 139–152. [CrossRef]

38. De Livera, A.M.; Hyndman, R.J.; Snyder, R.D. Forecasting Time Series With Complex Seasonal Patterns Using Exponential Smoothing. *J. Am. Stat. Assoc.* **2011**, *106*, 1513–1527. [CrossRef]

39. Shu, Y.; Jin, Z.; Zhang, L.; Wang, L.; Yang, O.W. Traffic prediction using FARIMA models. In Proceedings of the 1999 IEEE International Conference on Communications (Cat. No. 99CH36311), Vancouver, BC, Canada, 6–10 June 1999.

40. Bell, R.W.; Hillmer, S.C. Modeling time series with calendar variation. *J. Am. Stat. Assoc.* **1983**, *78*, 526–534. [CrossRef]

41. Beale, H.D.; Demuth, H.B.; Hagan, M.T. *Neural Network Design, Martin Hagan*; Pws: Boston, MA, USA, 2014.

42. Korea Electric Power Corporation. Available online: http://www.kepco.co.kr (accessed on 9 May 2020).

Article

Partial Discharge Data Matching Method for GIS Case-Based Reasoning

Jiejie Dai [1,*], Yingbing Teng [1], Zhaoqi Zhang [2], Zhongmin Yu [3], Gehao Sheng [2] and Xiuchen Jiang [2]

[1] State Grid Shanghai Shinan Electric Power Supply Company, Xinbei Road No.268, Shanghai 201199, China
[2] Department of Electrical Engineering, Shanghai Jiao Tong University, Dongchuan Road No.800, Shanghai 200240, China
[3] State Grid Shanghai Electric Power Company, Yuanshen Road No.1122, Shanghai 200120, China
* Correspondence: secess@163.com

Received: 6 August 2019; Accepted: 24 September 2019; Published: 26 September 2019

Abstract: With the accumulation of partial discharge (PD) detection data from substation, case-based reasoning (CBR), which computes the match degree between detected PD data and historical case data provides new ideas for the interpretation and evaluation of partial discharge data. Aiming at the problem of partial discharge data matching, this paper proposes a data matching method based on a variational autoencoder (VAE). A VAE network model for partial discharge data is constructed to extract the deep eigenvalues. Cosine distance is then used to calculate the match degree between different partial discharge data. To verify the advantages of the proposed method, a partial discharge dataset was established through a partial discharge experiment and live detections on substation site. The proposed method was compared with other feature extraction methods and matching methods including statistical features, deep belief networks (DBN), deep convolutional neural networks (CNN), Euclidean distances, and correlation coefficients. The experimental results show that the cosine distance match degree based on the VAE feature vector can effectively detect similar partial discharge data compared with other data matching methods.

Keywords: partial discharge; gas insulated switchgear; case-based reasoning; data matching; variational autoencoder

---

1. Introduction

CIGRE's statistics show that about 30% dielectric failures of gas insulated switchgears (GIS) are related to design deficiencies [1]. Through the analysis of a large amount of partial discharge (PD) data from GIS in service, we also found that the proportion of PD cases caused by design reasons is high. This leads to a situation that the same type GIS equipment from the same manufacturer are susceptible to repeat partial discharge on similar location. This provides the basis for case-based reasoning (CBR) in GIS. Case-based reasoning is a branch of artificial intelligence (AI) that provides answers to new questions based on experience in historical cases [2,3]. In the latest studies, CBR has been used in load forecasting, energy management, grid system safety assessment, and power equipment failure assessment [4–7]. In Reference [8], a case-based reasoning method is utilized to diagnose the incipient fault of power transformer. Pretreated dissolved gas analysis (DGA) data is used in the CBR system. Reference [9] developed a case-based reasoning approach for identifying and filtering acoustic emission (AE) noise signals. The paper proposed a parametric case representation method for the AE signal process. Since CBR requires the accumulation of cases and data in the early stage, there is no CBR related literature published in the field of partial discharge. After accumulating a large amount of GIS PD detection data from substation site, CBR can provide new ideas for the interpretation and evaluation of partial discharge data. The key step in a CBR system is the case matching strategy.

PD data is one of the key features in a GIS PD detection case. So this paper focused on the data matching problem in the CBR system establishment. A structure of GIS, CBR used PD data, the match degree is presented and shown in Figure 1.

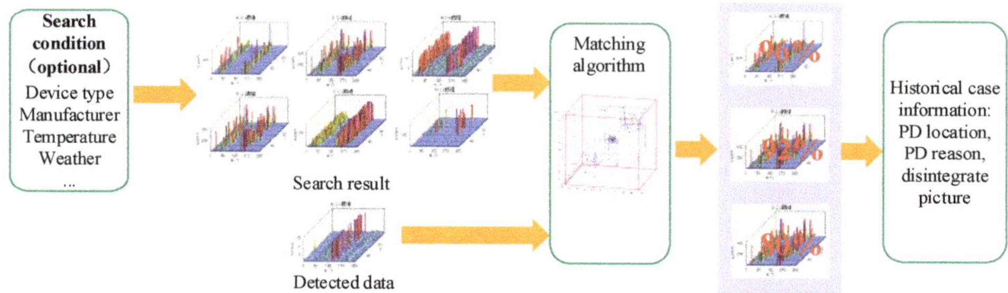

Figure 1. The framework of partial discharge (PD) data matching.

Some phase resolved pulse sequence (PRPS) graphs are used in Figure 1 to refer to the data detected by the GIS partial discharge ultra-high frequency (UHF) detection. The specific procedures are as follows: First, the historical data are retrieved from the historical case database according to the operating conditions of the detected equipment, the manufacturer and other search conditions; the detected data are then matched with the historical data, and those cases for which the data match degree exceeds a threshold are considered match cases; and finally, from the match case, we can obtain information such as the highest probability of PD location in the detected power equipment, the most likely cause of PD in the detected power equipment, and pictures of disintegrated power equipment in historical cases. Maintenance plans can be developed based on match information. Therefore, PD history detection data can be more effectively utilized and can provide a basis for data-driven device status evaluations.

There are two key processes that are used to calculate PD data match degree. The first key process is to extract the valid eigenvalues for PD data, and the second is to obtain the match degree (MD) based on the eigenvectors. The traditional feature extraction methods used for PD data extract a variety of statistical features from, for example, histograms, scatter plots, and grayscale images based on PRPD (phase resolved partial discharge) data [10–12]. Moreover, there are also some other algorithms applied to PD data feature extraction, such as principal component analysis (PCA) [13], wavelet packets transformation [14], sparse representation [15], and signal norms [16]. The algorithms proposed in the references behaved a good performance in the task of PD pattern recognition. However, due to the multi-source heterogeneity of access data in big data centers, the huge differences in the performances of PD detect instruments and the complex operating environments in substations, the statistical characteristics obtained by the traditional statistical methods have become inadequate in identifications of typical partial discharge types. In addition, data matching of PD data needs even more stringent requirements than those for PD pattern recognition.

In recent years, related technologies such as deep auto-encoders, deep convolutional networks, recurrent neural networks, and deep belief networks have shown good performance in many fields, including image processing and speech processing [17–21]. Reference [22] studied the application of deep neural networks in the diagnosis of partial discharges and demonstrated the improvements in accuracy and visualization that can be obtained through the deep learning method. Reference [23] obtained a two-dimensional spectral frame representation of a UHF signal employing a time-frequency analysis and then used a deep convolutional network to obtain enhanced features under different PD sources. Auto-encoding (AE) is an unsupervised feature learning method, and its hidden layer can effectively extract the internal expression of data. Its deep structure makes the network closer to the human brain's information hierarchical processing, with better nonlinear modelling ability [24,25].

The variational autoencoder (VAE) proposed by Kingma et al. is a generating network based on variational Bayesian inference [26]. It avoids the computational complexity of dataset likelihood probability calculations and traditional Monte Carlo sampling and is therefore becoming an area of considerable research interest in text classification, semi-supervised learning, and other related fields.

This paper presents a PD data matching method based on VAE. The network uses variational Bayesian method to quickly approximate the posterior probability and extract the deep features of PD data. Euclidean distance, cosine distance, and correlation coefficient (Cc) methods were used to measure the similarity between different data, the comparative results of which are also shown in this paper.

The rest of the paper is organized as follows. Section 2 introduces basic information on variational autoencoder networks. Section 3 provides further information on the proposed partial discharge data matching approach. The dataset used in this paper is described in Section 4. Section 5 validates the data matching approach with different case studies and discusses the results obtained. The conclusions are presented in Section 6.

2. Variational Autoencoder

Variational Bayes inference [27] is a deterministic approximation method that maximizes the lower bound of the marginal likelihood function of the observed data by iteratively updating the variational parameters and approximates the posterior probability of unobservable variables.

For a sample set X, define the eigenvalues of the data as latent variables z because they cannot be directly observed. According to the Bayesian criterion, the posterior probabilities of the latent variables z are

$$p(z|x) = \frac{p(x|z)p(z)}{p(x)} \tag{1}$$

It is difficult to obtain an exact analytical solution for $p(x)$, therefore, in the variational Bayes inference, an approximate distribution $q(z|x)$ is introduced to fit the real posterior distribution $p(z|x)$. Kullback-Leibler (KL) divergence is used to compare the similarities of the two distributions.

$$KL(p(z|x)\|q(z|x)) = \sum p(z|x) \log \frac{p(z|x)}{q(z|x)} \tag{2}$$

The approximate distribution $q(z|x)$ is estimated by an auto-encoder network in VAE. VAE consists of a probabilistic encoder and a probabilistic decoder and uses a stochastic gradient variational Bayes algorithm to achieve a posterior distribution model that optimizes the hidden layer.

According to the variational Bayes method, the log marginal likelihood of the sample data X can be simplified as shown below.

$$\log p_\theta(x^{(i)}) = KL(q_\phi(z|x^{(i)})\|p_\theta(z|x^{(i)})) + \mathcal{L}(\theta, \phi; x^{(i)}) \tag{3}$$

where ϕ is the real posterior distribution parameter, and θ is the approximate distribution parameter of the hidden layer. The first item is the KL divergence between the approximate distribution of the hidden layer and the real posterior distribution. Since KL divergence is nonnegative, the KL divergence is zero only if the two distributions are exactly the same [28]. Thus, $\log p_\theta(x^{(i)}) \geq \mathcal{L}(\theta, \phi; x^{(i)})$. Equation (3) can be expanded:

$$\log p_\theta(x^{(i)}) \geq \mathcal{L}(\theta, \phi; x^{(i)}) = -KL(q_\phi(z|x^{(i)})\|p_\theta(z)) + E_{q_\phi(z|x^{(i)})}[\log p_\theta(x^{(i)}|z)] \tag{4}$$

The optimal approximation of the sample set $p_\theta(x^{(i)})$ can be obtained by maximizing variational bound $L(\theta, \varphi; x(i))$ [29].

3. Data Matching Method of Partial Discharge Based on VAE

The encoder section of the VAE model for partial discharge data can be represented by Equation (5).

$$\begin{cases} z = \mu_{enc} + \sigma_{enc} \odot \varepsilon & \varepsilon \sim \mathcal{N}(0, I) \\ \mu_{enc} = f(W_{\mu_{enc}} h_1 + b_{\mu_{enc}}) \\ \log \sigma_{enc} = f(W_{\sigma_{enc}} h_1 + b_{\sigma_{enc}}) \\ h_1 = f(W_{h_1} x + b_{h_1}) \end{cases} \tag{5}$$

where W and b are the weights and biases of each layer, and x is the input vector. h_1, μ_{enc}, and σ_{enc} are the outputs of the first and second layers of the network. f is the activation function. Based on Gaussian distribution parameters μ and σ, the hidden layer output z is obtained by sampling $q(z|x(i))$, and $N(0,I)$ is the standard normal distribution.

The decoder section of the VAE model for partial discharge data can be represented by the following equation.

$$\begin{cases} \log p(x|z) = \log \mathcal{N}(x; \mu_{dec}, \sigma_{dec}^2 I) \\ \mu_{dec} = f(W_{\mu_{dec}} h_2 + b_{\mu_{dec}}) \\ \log \sigma_{dec} = f(W_{\sigma_{dec}} h_2 + b_{\sigma_{dec}}) \\ h_2 = f(W_{h_2} z + b_{h_2}) \end{cases} \tag{6}$$

where W and b are the weights and biases of each layer, h_2, μ_{dec}, and σ_{dec} are the outputs of each layer of the decoder, and f is the activation function.

The target optimization function of Equation (4) can be rewritten as Equation (7).

$$\mathcal{L}(\theta; x^{(i)}) = \frac{1}{2} \sum_{j=1}^{J} \left(1 + \log((\sigma_{encj}^{(i)})^2) - (\mu_{encj}^{(i)})^2 - (\sigma_{encj}^{(i)})^2\right) + \frac{1}{L} \sum_{l=1}^{L} (\log p_\theta(x^{(i)}|z^{(i,l)})) \tag{7}$$

where J is the dimension of the latent variables z, and L is the number of samples of the latent variables z on the posterior distribution.

The parameters of the probability encoder and the probability decoder are then optimized by the stochastic gradient descent algorithm. When Equation (7) converges or stabilizes, the output of the encoder part of VAE is the extracted eigenvalues.

Figure 2 shows the matching process of partial discharge data based on the VAE model.

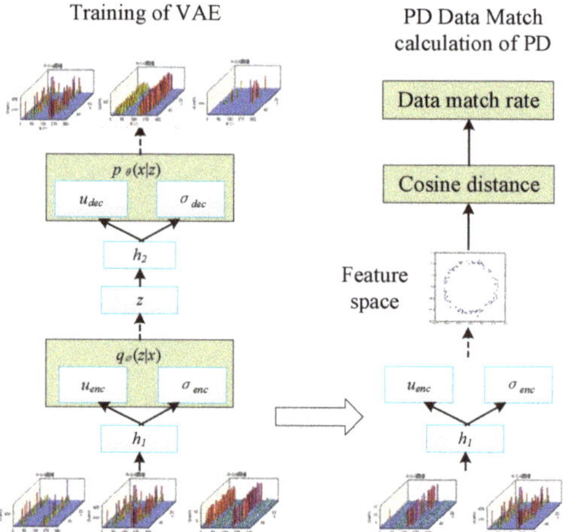

Figure 2. The procedure of feature extraction and match degree computation based on variational autoencoder (VAE).

The match degree of partial discharge data can be obtained by calculating the distance between the partial discharge data by using the cosine algorithm of Equation (8).

$$MD = \frac{V_a \cdot V_b}{\|V_a\| \times \|V_b\|} \tag{8}$$

where V_a and V_b are the eigenvectors extracted from the two PD datasets. $\|\bullet\|$ is the length of the vector.

4. Dataset

For this paper, the PD data sample sets were set up by laboratory partial discharge simulation and substation partial discharge live detection. We used the ultra-high frequency detection method and PRPS data format that is commonly used in UHF partial discharge detection. The dataset contained more than 20,000 pieces of simulated experimental data and more than 20,000 pieces of field test data.

4.1. Laboratory Experiment

Four typical partial discharge defect models were designed, and the experiment was conducted on a real GIS platform, the experimental connection diagram of which is shown in Figure 3. Typical design defects include floating electrode defects, metallic protrusion defects, insulation void discharge defects, and free metal particle discharge defects.

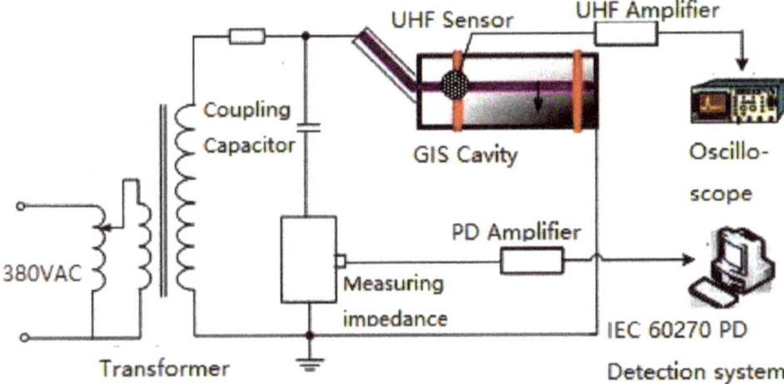

Figure 3. PD experiment circuit on a true gas insulated switchgears (GIS) model.

(1) Floating electrode defect: Epoxy resin was used to cast copper sheets of different sizes. The amount of discharge can be controlled by changing different epoxy blocks, as shown in Figure 4a.

(2) Metallic protrusion defect: The high-voltage terminal is connected to an aluminum tip electrode, and the ground terminal is connected with a Ø54 mm aluminum disc. By adjusting the size of the tip electrode and the height of the air gap between it and the ground electrode, it is possible to control the amount of discharge, as shown in Figure 4b.

(3) Particle discharge: The high-voltage terminal is connected to a ball electrode, and the low-voltage ground terminal is connected to a concave disk electrode, with free metal particles of different sizes and numbers placed in the center of it, as shown in Figure 4c.

(4) Insulation discharge: Casting the epoxy into a cylinder will leave bubbles of different sizes inside during the casting process, as shown in Figure 4d.

The nominal voltage of the GIS used in partial discharge experiment is 145 kV, the output voltage of experimental power supply is 0–220 kV. Typical partial discharge inception voltage (PDIV) and partial discharge extinction voltage (PDEV) for each type of defect are listed in Table 1. The main parameters of instrument used in the experiment are shown in Table 2.

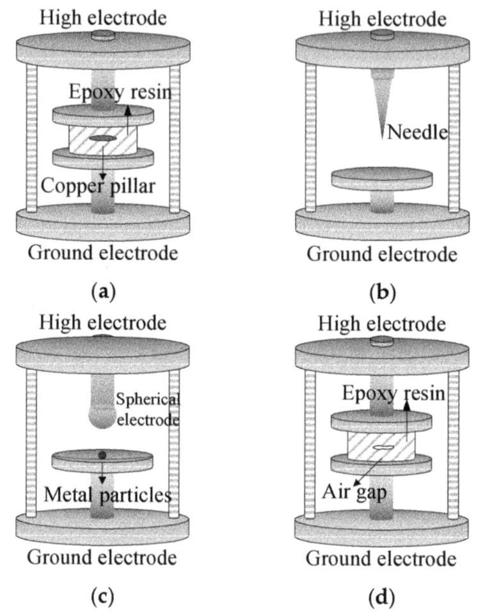

Figure 4. PD defect models. (**a**) Floating electrode discharge; (**b**) metallic protrusion discharge; (**c**) free metal particle discharge; (**d**) insulation void discharge.

Table 1. Typical partial discharge inception voltage (PDIV) and partial discharge extinction voltage (PDEV) in the experiment.

PD Type	PDIV (kV)	PDEV (kV)
Floating electrode discharge	128	112
Metallic protrusion discharge	67	60
Insulation void discharge	110	102
Free metal partial discharge	84	70

Table 2. The main parameters of instrument used in the experiment.

Instrument Type	Key Parameter
IEC60270 Digital partial discharge detector	Sensitivity: <0.1 pC System bandwidth: 30 kHz–1.5 MHz
Oscilloscope	Sampling rate: single channel 10 GS/s Analog bandwidth: 2 GHz

The typical PRPS data detected in the simulation experiment were normalized, as shown in Figure 5.

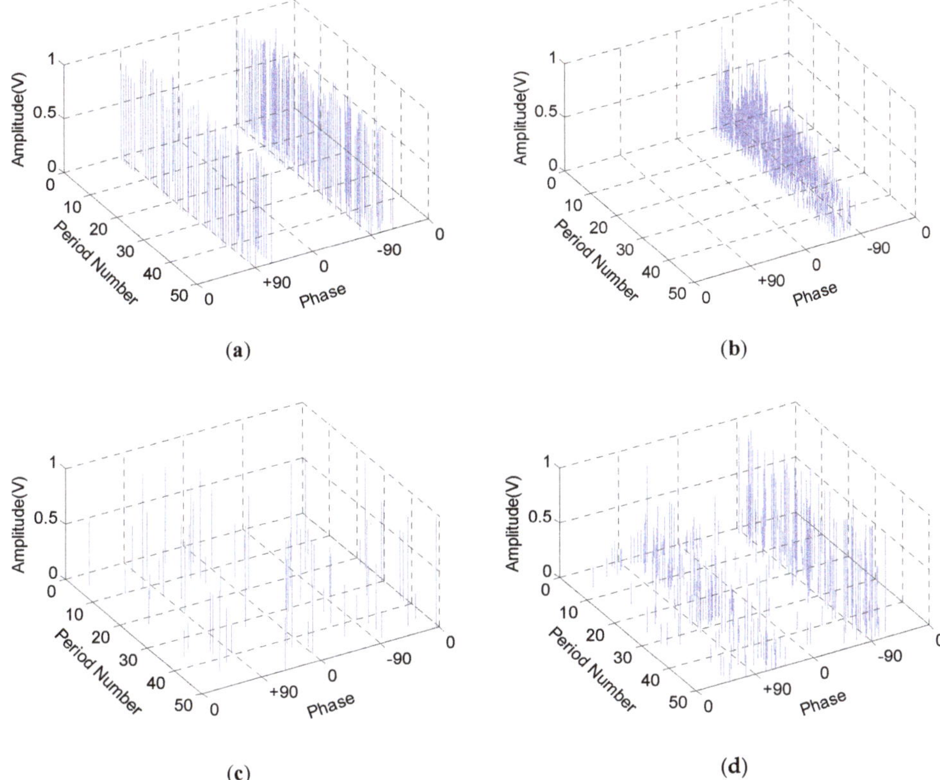

Figure 5. Typical PD phase resolved pulse sequence (PRPS) data from a true GIS model experiment. (**a**) Floating electrode discharge; (**b**) metallic protrusion discharge; (**c**) free metal particle discharge; (**d**) insulation void discharge.

4.2. Substation On-Site Detection

In the past five years, we have accumulated a large amount of on-site detection data by periodically conducting PD tests for more than 30 substations in China. Among those data, there are 42 cases in which the power equipment defects have been verified by disassembly overhaul, including floating electrode defects, metallic protrusion defects, insulation void discharge defects, and free metal particle discharge defects. The statistical information related to the cases is shown in Table 3.

As seen from Table 1, in all disintegrative cases, the proportion of similar cases for the same PD type was high. Therefore, for some detected data from equipment suspected of being defective, there is a high probability that similar data can be found from its historical cases, especially for those data that can be recognized as floating and insulation discharges.

Table 3. Information of on-site power equipment disintegration verification cases.

PD Type	Number of Cases	Defect Reason	Number of Cases
Floating Electrode Discharge	24	Poor contact in disconnector	6
		Loose connection in equipotential leaf spring	10
		The build in sensors is not effectively grounded	5
		Other reasons	3
Metallic Protrusion Discharge	2	Quality defects in conductor	2
Insulation Void Discharge	14	Quality defects in supporting insulator	4
		Aging of insulation	3
		Installation defects	2
		Other reasons	5
Free Metal Particle Discharge	2	Installation defects	2

5. Experiment and Results Analysis

5.1. Experiment Setup

The main flow of the comparative experiment is shown in Figure 6.

First, the training set was composed of both laboratory experimental data and substation field detection data. The experimental data and the field detection data were mixed and disordered. An unsupervised training was performed to the established VAE model on this dataset, to obtain a feature extraction model with better generalization performance. The test set consisted of only substation field detection case data which the defect is verified by GIS disintegration. The data from four GIS disintegration cases were selected in order to examine the matching performance of the data matching model for case data. These four cases contained different similar situations. The matching degrees were calculated between data in four cases, and the different feature extraction methods and different matching degree calculation methods were compared. Finally, the generalization capabilities of different methods were analyzed on all the 42 cases.

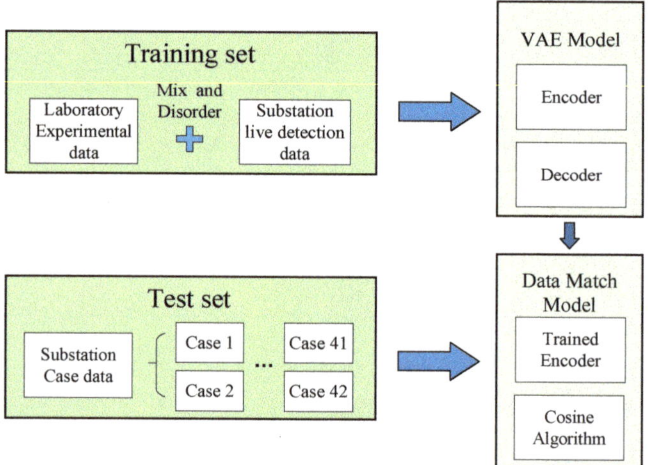

Figure 6. Experimental flow chart.

The baseline systems that were used for feature extraction are now briefly described.

(1) Statistical eigenvalues: They are a commonly used feature extraction method in PD data processing. The traditional statistical eigenvalues consist of 16 characteristic parameters such as skewness (Sk), steepness (Ku), asymmetry (Q), the cross correlation coefficient (Cc) of the PD amplitude, and PD numbers in the positive and negative half of the power frequency cycle [30].

(2) DBN: A deep belief network (DBN) consists of multi-layer RBMs. The DBN network used for comparison had six layers, and the numbers of units for each layer were 3600, 1000, 500, 100, 10, and 4. In addition, the output of the second to last layer is used as the extracted eigenvalue. The detailed calculation can be seen in Reference [31].

(3) CNN: A deep convolutional neural network (CNN) consists of a number of two-dimensional convolutional kernels and uses multi-layer convolutional and pooling operations to obtain deep features of data. The CNN input layer used for this paper was 50×72, the two convolutional layers were six convolutional kernels of 3×3 and 36 convolutional kernels of 3×3, and the corresponding pooling layers were 1×2 and 1×11. The numbers of two fully connected layers were 500 and 10, and the number of output layers was four. The input of the output layer was used as the extracted eigenvalue. Detailed calculations can be seen in Reference [23].

The baseline systems used for the MD calculation are now briefly described.

(1) Euclidean distance MD: MD is obtained based on the Euclidean distance [32] between two groups of vectors. The problem with matches based on Euclidean distance is that it is difficult to determine the appropriate standard, and thus normalization is difficult. For this paper, the maximum distance in all sample data was selected as the standard, and MD was calculated according to the following formula:

$$MD = 1 - \frac{D_{ab}}{D_{\max}} \tag{9}$$

where D_{ab} is the Euclidean distance between the eigenvectors extracted from the two PD data. $D_{\max}$ is the maximum Euclidean distance between the PD data in the dataset.

(2) Correlation coefficient: A correlation coefficient is a measure of the linear correlation between two variables [33]. It has a value between +1 and −1, where one is total positive linear correlation, zero is no linear correlation, and −1 is total negative linear correlation. Therefore, the MD can be calculated according to the following formula:

$$MD = |r_{ab}| \tag{10}$$

where r_{ab} is the correlation coefficient between the eigenvectors extracted respectively from the two PD data.

The experimental platform was configured as a Core i7 processor (Intel, Santa Clara, CA, USA) operating at 3.9 GHz with 16 GB of memory, the operating system was Ubuntu 14.0 (Canonical, London, UK), and the code was implemented in Python. For the results presented in this paper, the dimension of each PD data was 50×72. The VAE used in the study consisted of seven layers: An input layer, an output layer, a latent layer, and four intermediate layers. The structure of the network is shown in Table 4.

Table 4. VAE model parameters for partial discharge data feature extraction.

Layer Number	Layer Type	Number of Neurons	Activation Function
1	Input layer	3600	-
2	Hidden layer	1000	ReLU
3	Hidden layer	500	ReLU
4	Latent variables layer	2	Gaussian distribution
5	Hidden layer	500	ReLU
6	Hidden layer	1000	ReLU
7	Output layer	3600	Sigmoid

Network layers 1–4 formed the encoder part of VAE, and layers 4–7 formed the decoder part of VAE. The output of the latent variables layer was the extracted eigenvalues. Using the established PD dataset, the VAE was trained without supervision, as described in Section 3.

5.2. The Comparison between Different Feature Extraction Methods

We selected four cases of partial discharge detection verified by disintegration and numbered them cases 1–4. The case information is shown in Table 5.

The equipment in Case 1 and Case 2 belonged to the same manufacturer and were of the same type. They also had the same discharge location. Case 3 has the same PD pattern as for Cases 1 and 2, but the equipment manufacturers and discharge locations differed. Case 4 was a comparative case with different PD types. The partial discharge data detected in the above four cases are shown in Figure 7.

Table 5. Information for four partial discharge detection substation site cases.

Case Number	PD Type	PD Location
1	Floating discharge	Joint of insulation tension pole and transmission gear in disconnector
2	Floating discharge	Joint of insulation tension pole and transmission gear in disconnector
3	Floating discharge	Built-in sensor connector
4	Insulation discharge	Cable terminal damaged in GIS

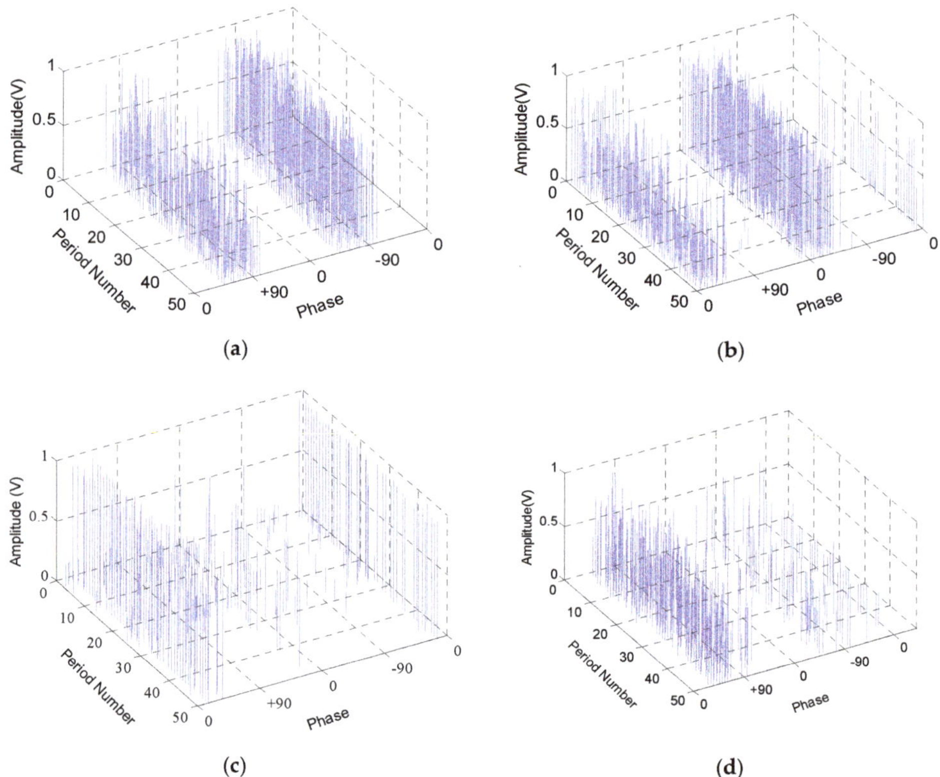

Figure 7. Data from four partial discharge detection substation site cases. (**a**) Case 1; (**b**) case 2; (**c**) case 3; (**d**) case 4.

The trained VAE network model was used to extract the features of the partial discharge data in the above four cases and to calculate the MD between them. At the same time, the statistical characteristics, DBN eigenvalues, and CNN eigenvalues for the four cases data were used to calculate MD. All the MDs were based on cosine distance. The results are shown in Table 6.

Table 6. Match degree for the different feature extraction methods between data from four on-site detection cases.

Case Number	VAE	Statistical	DBN	CNN
1-2	96.87%	66.42%	90.41%	88.62%
1-3	73.78%	59.33%	85.38%	88.46%
1-4	6.93%	40.41%	9.28%	7.41%
2-3	61.75%	60.27%	86.50%	89.15%
2-4	1.92%	39.53%	5.74%	6.46%
3-4	9.51%	26.98%	11.72%	11.96%

The information of four cases in Table 6 are described in Table 5. The similarity between case 1 and case 2 should be 100% because they have GIS devices produced by the same manufacturer, and partial discharge occurs at the same position, so the similarity result should definitely be the higher the better. Case 3 has the same PD type compared to cases 1, 2, but the case details such as PD location and reason are different. Case 4 has a completely different PD type, and the similarity should be 0%, so the similarity result should definitely be the smaller the better. As seen from Table 6, using the VAE method, case 1-2 had a higher MD than those of the other cases and was 23.09% higher than that of case 1-3 and 89.94% higher than that of case 1-4. As a comparison, for the MD results based on statistical eigenvalues, case 1-2 was 7.09% higher than case 1-3 and 26.01% higher than case 1-4, which means that the MD based on statistical eigenvalues were relatively close. It had a lower distinguishing ability, even for different PD pattern recognitions. Regarding the MD results based on DBN and CNN, the MD of case 1-4 and case 2-4 were obviously lower than those of case 1-2 and case 1-3. Therefore, the DBN and CNN models performed better for data from different PD types than the traditional statistical method. However, for cases 1-2, 1-3, and 2-3, the MDs were too close to distinguish similar and dissimilar cases and were therefore less effective than the VAE model.

5.3. The Comparison between Different Match Degree Calculation Methods

To investigate the effects of the different match degree calculation methods, the MD were calculated by cosine distance, Euclidean distance, and correlation coefficient, respectively based on the VAE eigenvalues for the four cases in Table 5. The results are shown in Table 7.

Table 7. Comparison of different matching calculation methods on four detection cases on-site.

Case Number	Cosine Distance	Euclidean Distance	Correlation Coefficient
1-2	96.87%	93.03%	95.70%
1-3	73.78%	75.23%	77.64%
1-4	6.93%	8.31%	12.67%
2-3	61.75%	60.84%	79.14%
2-4	1.92%	6.80%	18.22%
3-4	9.51%	7.18%	13.95%

It can be seen that there were slight differences in the specific values among the methods, but overall, all the methods had good ability to distinguish between similar and dissimilar cases. Furthermore, we calculated the MDs on all the data for the 42 cases. The VAE model was used to extract the eigenvalues, and the MD were calculated by cosine distance, Euclidean distance, and correlation coefficient, respectively. The match result is defined accurate if the MD exceeds 80% under the similar cases and less than 20% under the dissimilar cases. The accuracies are shown in Table 8.

Table 8. Comparison of the different matching calculation methods for all the on-site cases.

PD Type	Floating Discharge	Metallic Protrusion Discharge	Insulation Discharge	Particle Discharge
Cosine distance	82.6%	79.3%	85.7%	89.6%
Euclidean distance	63.6%	53.5%	75.3%	40.4%
Correlation coefficient	77.5%	48.5%	70.9%	39.9%

It can be seen from Table 8 that for a large number of cases, the accuracy of MD based on Euclidean distance and the correlation coefficient were lower than those based on the cosine distance. The reason was that in the calculation of the Euclidean distance MD, data from all kinds of cases were compared with the fixed maximum distance, thus the singular value will result in the poor effect overall. In the calculation of MD based on correlation coefficient, more MD values exceeded 20% under dissimilar cases. In addition, because a large amount of PD data was stored in each case, there may have been some low quality data that differed greatly from other data in the same case. To improve performance in big data engineering applications, the data cleaning method needs to be used for data filtration in future research.

5.4. The Comparison between Different Threshold

The different definition of accurate match also has an important impact on the final effect of the CBR system. In Section 5.3, the match result is defined accurate if the MD exceeds 80% under the similar cases and less than 20% under the dissimilar cases. The accuracies change with the threshold changes. In the classification problem, the number 50% is usually used as the threshold for the classification output. If the output of a category is greater than 50%, the sample can be classified into this category. If it is less than 50%, it is not considered to be the category. In data matching applications, it is necessary to adopt differentiated thresholds to get more accurate case results. However, different algorithms have different adaptability to different threshold settings. To investigate the accuracy of different algorithms at different thresholds, we performed the following experiments.

Firstly, 1000 sets of data were selected from similar cases, and the eigenvalues of the data in each case were calculated by VAE, statistical eigenvalue, DBN and CNN, and the MDs were obtained by the cosine algorithm. The thresholds were defined as 50%, 60%, 70%, 80%, and 90%, respectively. The match result is defined accurate if the MD exceeds the threshold, and the accuracy obtained by different algorithms is shown in Figure 8a.

Secondly, 1000 sets of data were selected from dissimilar cases, and the eigenvalues of the data in each case were calculated by VAE, statistical eigenvalue, DBN and CNN, and the MDs were obtained by the cosine algorithm. The thresholds were defined as 50%, 40%, 30%, 20%, and 10%, respectively. The match result is defined accurate if the MD less than the threshold, and the accuracy obtained by different algorithms is shown in Figure 8b.

As seen from Figure 8a, in the comparative analysis under similar case data, when the threshold was set to 70% or less, the accuracy obtained by CNN, DBN, and VAE has a small difference. When the threshold was set to 80% or more, the accuracy of CNN and DBN decreased more obviously, while the accuracy of VAE can still reach more than 60% at the threshold of 90%. The reason is that under similar cases, the MDs obtained by VAE were mainly distributed above 0.9, while the MDs calculated by CNN and DBN were mainly distributed between 0.7 and 0.8. The MDs calculated by the statistical eigenvalues were mainly distributed between 0.5 and 0.6. in the comparative analysis under dissimilar case data, when the threshold was set to 40% or more, the accuracy obtained by the four methods was not much different. When the threshold was set to 20% or less, the accuracy of the four methods all reduced, while the accuracy of VAE can still reach more than 40% at the threshold of 10%. Under dissimilar cases, the MDs obtained by VAE were mainly distributed below 0.2, while the MDs calculated by CNN and DBN were mainly distributed between 0.1 and 0.4. The MDs calculated by the statistical eigenvalues were mainly distributed between 0.3 and 0.4. Taken together, the VAE model also has better generalization capabilities for threshold changes.

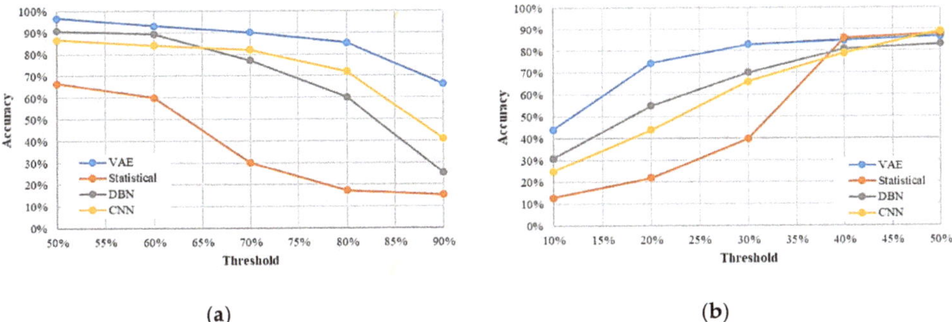

(a) (b)

Figure 8. Accuracy of the different methods. (**a**) Accuracy of the different methods under similar cases; (**b**) accuracy of the different methods under dissimilar cases.

6. Conclusions

This paper has proposed a PD data matching method based on a VAE network to perform data mining on historical PD databases. Similar cases found by the method can provide abundant information for PD diagnosis and equipment status evaluation. A PD dataset was established from a laboratory partial discharge experiment and substation live detections. Additionally, on the data set, a comparative experiment was conducted on the VAE and the comparison method. Experimental results show:

(1) Compared with traditional statistical eigenvalues, deep learning related methods, such as CNN, DBN, VAE, etc., have better effects on the identification of different PD types on complex data sets;

(2) Compared with CNN, the DBN and VBE models extracted the partial discharge data eigenvalues with better expression ability. In the data matching experiment, the discrimination degree is higher.

(3) The MD calculation method of cosine distance has better precision under a large number of samples than the Euclidean distance and correlation coefficient.

The work in this paper provides a new way of thinking about PD data mining under the background of big data. In further research, a better match strategy will be designed to meet the engineering requirements of PD data mining. The benchmarking criteria for MD of PD data is the key issue to be studied in the next step.

Author Contributions: Data curation, J.D.; Formal analysis, G.S.; Funding acquisition, X.J.; Methodology, J.D.; Resources, J.D.; Software, Z.Z.; Supervision, Z.Y.; Validation, Z.Z. and Y.T.; Writing—original draft, J.D.; Writing—review and editing, Z.Z. and Y.T.

Funding: This work was supported in part by the National Key Research and Development Program of China (Grant ID. 2017YFB0902705), as well as the Science and Technology Project of State Grid Corporation in China.

Conflicts of Interest: The authors declare no conflict of interest.

References

1. Schichler, U.; Koltunowicz, W.; Endo, F.; Feser, K.; Giboulet, A.; Girodet, A.; Girodet, A.; Hama, H.; Hampton, B.; Kranz, H.G.; et al. Risk assessment on defects in GIS based on PD diagnostics. *IEEE Trans. Dielectr. Electr. Insul.* **2013**, *20*, 2165–2172. [CrossRef]

2. Kang, Y.B.; Krishnaswamy, S.; Zaslavsky, A. A retrieval strategy for case-based reasoning using similarity and association knowledge. *IEEE Trans. Cybern.* **2014**, *44*, 473–478. [CrossRef] [PubMed]

3. Jeong, M.; Morrison, J.R.; Suh, H. Approximate life cycle assessment via case-based reasoning for eco-design. *IEEE Trans. Autom. Sci. Eng.* **2015**, *12*, 716–728. [CrossRef]

4. Platon, R.; Dehkordi, V.R.; Martel, J. Hourly prediction of a building's electricity consumption using case-based reasoning, artificial neural networks and principal component analysis. *Energy Build.* **2015**, *92*, 10–18. [CrossRef]

5. De la Rosa, J.J.G.; Agüera-Pérez, A.; Palomares-Salas, J.C.; Sierra-Fernández, J.M.; Moreno-Muñoz, A. A novel virtual instrument for power quality surveillance based in higher-order statistics and case-based reasoning. *Measurement* **2012**, *45*, 1824–1835. [CrossRef]

6. Nandanwar, S.R.; Warkad, S.B. Application of case based reasoning in voltage security assessment. In Proceedings of the Computational Intelligence on Power, Energy and Controls with their Impact on Humanity (CIPECH), Ghaziabad, India, 18–19 November 2016; pp. 19–22.

7. Tiako, R.; Jayaweera, D.; Islam, S. Real-time dynamic security assessment of power systems with large amount of wind power using Case-Based Reasoning methodology. In Proceedings of the Power and Energy Society General Meeting (PESGM), San Diego, CA, USA, 22–26 July 2012; pp. 1–7.

8. Qian, Z.; Gao, W.S.; Yan, Z. Application of case-based reasoning theory in fault diagnosis of power transformer. In Proceedings of the 8th International Conference on Properties and Applications of Dielectric Materials, Bali, Indonesia, 26–30 June 2006; pp. 242–245.

9. Alekhin, R.; Varshavsky, P.; Eremeev, A.; Kozhevnikov, A. Application of the case-based reasoning approach for identification of acoustic-emission control signals of complex technical objects. In Proceedings of the 2018 3rd Russian-Pacific Conference on Computer Technology and Applications (RPC), Vladivostok, Russia, 18–25 August 2018; pp. 1–4.

10. Truong, L.H.; Lewin, P.L. Phase resolved analysis of partial discharges in liquid nitrogen under AC voltages. *IEEE Trans. Dielectr. Electr. Insul.* **2013**, *20*, 2179–2187. [CrossRef]

11. Florkowski, M.; Florkowska, B. Phase-resolved rise-time-based discrimination of partial discharges. *IET Gener. Transm. Distrib.* **2009**, *3*, 115–124. [CrossRef]

12. Sahoo, N.C.; Salama, M.M.A.; Bartnikas, R. Trends in partial discharge pattern classification: A survey. *IEEE Trans. Dielectr. Electr. Insul.* **2005**, *12*, 248–264. [CrossRef]

13. Wong, J.K.; Illias, H.A.; Bakar, A.H.A. A novel high noise tolerance feature extraction for partial discharge classification in XLPE cable joints. *IEEE Trans. Dielectr. Electr. Insul.* **2017**, *24*, 66–74.

14. Evagorou, D.; Kyprianous, A.; Lewin, P.; Stavrou, A.; Efthymiou, V.; Metaxas, A.C.; Georghiou, G.E. Feature extraction of partial discharge signals using the wavelet packet transform and classification with a probabilistic neural network. *IET Sci. Meas. Technol.* **2010**, *4*, 177–192. [CrossRef]

15. Majidi, M.; Fadali, M.S.; Etezadi-Amoli, M.; Oskuoee, M. Partial discharge pattern recognition via sparse representation and ANN. *IEEE Trans. Dielectr. Electr. Insul.* **2015**, *22*, 1061–1070. [CrossRef]

16. Majidi, M.; Oskuoee, M. Improving pattern recognition accuracy of partial discharges by new data preprocessing methods. *Electr. Power Syst. Res.* **2015**, *119*, 100–110. [CrossRef]

17. LeCun, Y.; Bengio, Y.; Hinton, G. Deep learning. *Science* **2015**, *521*, 436–444. [CrossRef] [PubMed]

18. Hinton, G.E.; Salakhutdinov, R.R. Reducing the dimensionality of data with neural networks. *Science* **2006**, *313*, 504–507. [CrossRef] [PubMed]

19. Kappeler, A.; Yoo, S.; Dai, Q.; Katsaggelos, A.K. Video super-resolution with convolutional neural networks. *IEEE Trans. Comput. Imaging* **2016**, *2*, 109–122. [CrossRef]

20. Garcia, C.; Delakis, M. Convolutional face finder: A neural architecture for fast and robust face detection. *IEEE Trans. Patt. Anal. Mach. Int.* **2004**, *26*, 1408–1423. [CrossRef]

21. Lecun, Y.; Bottou, L.; Bengio, Y.; Haffner, P. Gradient-based learning applied to document recognition. *Proc. IEEE* **1998**, *86*, 2278–2324. [CrossRef]

22. Catterson, V.M.; Sheng, B. Deep neural networks for understanding and diagnosing partial discharge data. In Proceedings of the IEEE Electrical Insulation Conference, Seattle, WA, USA, 7–10 June 2015; pp. 218–221.

23. Li, G.; Rong, M.; Wang, X.; Li, X.; Li, Y. Partial discharge patterns recognition with deep Convolutional Neural Networks. In Proceedings of the 2016 International Conference on Condition Monitoring and Diagnosis (CMD), Xi'an, China, 25–28 September 2016; pp. 324–327.

24. Chen, Z.; Li, W. Multisensor feature fusion for bearing fault diagnosis using sparse autoencoder and deep belief network. *IEEE Trans. Instrum. Meas.* **2017**, *66*, 1693–1702. [CrossRef]

25. Vincent, P.; Larochelle, H.; Lajoie, I.; Bengio, Y.; Manzagol, P. Stacked denoising autoencoders: Learning useful representations in a deep network with a local denoising criterion. *J. Mach. Learn. Res.* **2010**, *11*, 3371–3408.

26. Kingma, D.P.; Welling, M. Auto-Encoding Variational Bayes. Available online: https://arxiv.org/abs/1312.6114 (accessed on 20 December 2013).
27. Woolrich, M.W.; Behrens, T.E. Variational bayes inference of spatial mixture models for segmentation. *IEEE Trans. Med. Imaging* **2006**, *25*, 1380–1391. [CrossRef]
28. Tjandra, A.; Sakti, S.; Nakamura, S.; Adriani, M. Stochastic gradient variational bayes for deep learning-based ASR. In Proceedings of the IEEE Workshop on Automatic Speech Recognition and Understanding (ASRU), Scottsdale, AZ, USA, 13–17 December 2015; pp. 175–180.
29. Xu, W.; Sun, H.; Deng, C.; Tan, Y. Variational autoencoder for semi-supervised text classification. In Proceedings of the Thirty-First AAAI Conference on Artificial Intelligence and The Twenty-Ninth Innovative Applications of Artificial Intelligence Conference, San Francisco, CA, USA, 4–9 February 2017; pp. 3358–3364.
30. Li, L.; Tang, J.; Liu, Y. Partial discharge recognition in gas insulation switchgear based on multi-information fusion. *IEEE Trans. Dielectr. Electr. Insul.* **2015**, *22*, 1080–1087. [CrossRef]
31. Zhang, X.; Tang, J.; Pan, C.; Zhang, X.; Jin, M.; Yang, D.; Zheng, J.; Wang, T. Research of partial discharge recognition based on deep belief nets. *Power Syst. Technol.* **2016**, *40*, 3272–3278. (In Chinese)
32. Nguyen, C.; Lovering, C.; Neamtu, R. Ranked time series matching by interleaving similarity distances. In Proceedings of the IEEE International Conference on Big Data (Big Data), Boston, MA, USA, 1–14 December 2017; pp. 3530–3539.
33. Wang, X.; Song, G.; Chang, Z.; Luo, J.; Gao, J.; Wei, X.; Wei, Y. Faulty feeder detection based on mixed atom dictionary and energy spectrum energy for distribution network. *IET Gener. Transm. Distrib.* **2018**, *12*, 596–606. [CrossRef]

Article

Wind Farm NWP Data Preprocessing Method Based on t-SNE

Jiu Gu [1], Yining Wang [1], Da Xie [1,*] and Yu Zhang [2]

[1] School of Electronic Information and Electrical Engineering, Shanghai Jiao Tong University, Minhang District, Shanghai 200240, China; sjtugujiu@sjtu.edu.cn (J.G.); wangyining531@gmail.com (Y.W.)

[2] State Grid Shanghai Municipal Electric Power Company, Shanghai 200122, China; zhangyu@sh.sgcc.com.cn

*   Correspondence: profxzg@hotmail.com; Tel.: +86-21-34204298

Received: 20 August 2019; Accepted: 12 September 2019; Published: 23 September 2019

Abstract: The operation prediction of wind farms will be accompanied by the need for massive data processing, especially the preprocessing of wind farm meteorological data or numerical weather prediction (NWP). Because NWP data are strongly correlated with wind farm operation, proper processing of NWP data could not only reduce data volume but also improve the correlations of wind farm operation predictions. For this purpose, this paper proposes a data preprocessing algorithm based on t-distributed stochastic neighbor embedding (t-SNE). Firstly, the data collected were normalized to eliminate the influence caused by different dimensions. The t-SNE algorithm is then used to reduce the dimensionality of the NWP data related to wind farm operation. Finally, the wind farm data visualization platform is established. In this paper, 22 index variables in NWP data were taken as objects. The t-SNE method was used to preprocess the NWP historical data of a wind farm, and the results were compared with the results of the principal component analysis (PCA) algorithm. It outperformed PCA in error precision; in addition, t-SNE dimension reduction preprocessing also had a visual effect, which could be applied to big data visualization platforms. A long short-term memory network (LSTM) was used to predict the operation of the wind farm by combining the preprocessed NWP data and the operation data. The simulation results proved that the effect of the preprocessed NWP data based on t-SNE on the wind power prediction was significantly improved.

Keywords: t-SNE algorithm; numerical weather prediction; data preprocessing; data visualization; wind power generation

---

1. Introduction

Wind power is becoming one of the most important power sources in the power grid. At present, China's accumulated wind power capacity is 188 GW, and the total installed capacity has leapt to first in the world [1]. While the penetration rate of wind power is increasing, it generates a huge amount of data for recording the operational status of wind turbines, and so it needs to be studied using big data technology [2,3].

The key technologies of power big data include the following five parts: data acquisition, data storage, data preprocessing, data analysis, and data visualization. The five key parts of wind power big data technologies are shown in Figure 1. In the Figure 1, SCADA means supervisory control and data acquisition.

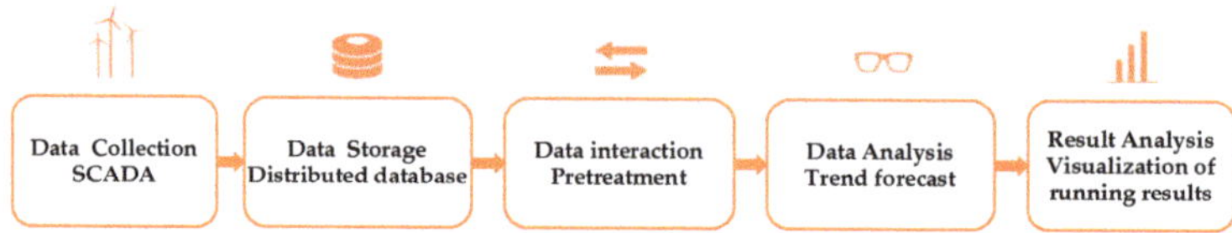

Figure 1. Important technologies of wind energy big data processing.

The acquisition and storage of data is the basis for an in-depth understanding of the operational status of wind turbines as contained in wind power big data; data preprocessing is a prerequisite for data analysis [4,5]; data analysis is the key to obtaining valuable information from massive data [6–9]; and data visualization is an intuitive and effective method of data presentation. Data preprocessing refers to the review, screening, sorting, transformation, statute, summary, and other processing done before the collected data is processed [10]. Unprocessed data obtained after data collection often have some problems. After preprocessing of data, it is possible to select and extract appropriate features for model training.

In terms of data preprocessing, Wang et al. [11] used data preprocessing techniques and swarm intelligence optimization algorithms to analyze wind speeds for wind energy potential assessment and prediction problems. Niu et al. [12] proposed a combined model for wind speed prediction, including a set of empirical mode decompositions of adaptive noise and a multi-target locust optimization algorithm. Jiang et al. [13] proposed a new hybrid model combining the de-drying method and an optimization algorithm with prediction technology for various unstable factors in complex power systems. Tian et al. [14] studied the accuracy of photovoltaic (PV) power prediction data, and proposed the processing of meteorological data by wavelet decomposition and principal component analysis. Malvoni et al. [15] proposed a cloud segmentation optimal entropy algorithm for the identification of unit anomaly data. Azimi et al. [16] proposed a new time-based K-means clustering method, including discrete wavelet transform, harmonic analysis, and multi-layer perceptual neural network methods, and developed a cluster selection method to determine the optimal training cluster. Zhao et al. [17] studied the feature reduction analysis of wind-induced anomaly data, and integrated the quadrilateral method and density-based clustering method to eliminate sparse outliers. Ye et al. [18] used the adjacent spatial correlation to establish an outlier identification algorithm based on the probabilistic wind farm power curve for the missing data problem in wind farm time series power data.

In terms of data analysis, Renani et al. [19] proposed a new backtracking algorithm for crossover and mutation operators for the problem of wind power prediction, and compared the advantages of an adaptive neuro-fuzzy inference system and other data mining algorithms. Zameer et al. [20] proposed a ML-STWP-based, machine-learning-based short-term wind energy prediction method for short-term wind power forecasting problems, and applied feature selection and regression learning techniques to wind power forecasting. Yuan et al. [21] proposed a hybrid model of the least squares support vector machine and gravity search algorithm for wind farm output power prediction. Abdoos et al. [22] used variational mode decomposition to decompose the time series for the wind power data prediction problem, and then used the Gram–Schmidt orthogonalization to eliminate the redundancy. Finally, the extreme learning machine algorithm was used to train the features.

The above research has mainly been aimed at the cleanup of bad data in wind power big datasets, and the recovery of missing data in the wind speed–power model. Atmospheric dynamics and detailed weather data such as wind direction, wind speed, atmospheric pressure, and air density also have important impacts on the operating state of wind farms, but they have not been paid much attention. Research on data reduction processing with such a large variety of data is also insufficient.

In this paper, a wind power data preprocessing method based on t-SNE has been proposed to reduce the dimensionality of the collected numerical weather prediction. The main work and problems of this paper are as follows:

(1) Applying the data dimensionality reduction algorithm of t-SNE to the preprocessing of numerical weather prediction (NWP) data of wind farms, and comparing this with the principal component analysis algorithm, the superiority of the algorithm was proven. A long short-term memory network (LSTM) network was used to predict the data after the dimension reduction using t-SNE and the original historical data, which proved that the method improves the prediction accuracy.

(2) Based on this, a wind farm visualization platform was established to display various types of data.

2. Preprocessing Algorithm

2.1. Normalized Processing

Assuming that the dataset has 2 dimensions, first calculate the influence of the zero mean difference and the covariance. The data after zero mean transformation is:

$$\begin{cases} x' = x - \bar{x} \\ y' = y - \bar{y} \end{cases} \tag{1}$$

The covariance of the new data is:

$$\sigma'_{xy} = \frac{1}{n}\sum_{i=1}^{n}\left(x'_i - \bar{x}'\right)\left(y'_i - \bar{y}'\right) \tag{2}$$

$\bar{x}' = 0, \bar{y}' = 0$, therefore:

$$\sigma'_{xy} = \frac{1}{n}\sum_{i=1}^{n}\left(x'_i\right)\left(y'_i\right) \tag{3}$$

The raw data covariance is:

$$\sigma_{xy} = \frac{1}{n}\sum_{i=1}^{n}(x_i - \bar{x})(y_i - \bar{y}) = \frac{1}{n}\sum_{i=1}^{n}\left(x'_i\right)\left(y'_i\right) \tag{4}$$

Therefore:

$$\sigma'_{xy} = \sigma_{xy} \tag{5}$$

After the variance is normalized, we have:

$$\begin{cases} x'' = \frac{x-\bar{x}}{\sigma_x} \\ y'' = \frac{y-\bar{y}}{\sigma_y} \end{cases} \tag{6}$$

After the variance is normalized, the covariance is as shown in Equation (7).

$$\sigma''_{xy} = \frac{1}{n}\sum_{i=1}^{n}\left(\frac{x-\bar{x}}{\sigma_x}\right)\left(\frac{y-\bar{y}}{\sigma_y}\right) = \frac{\sigma_{xy}}{\sigma_x\sigma_y} \tag{7}$$

The min–max normalization method is used for calculation, and the result of the linear function transformation is:

$$\begin{cases} x' = c_x \cdot x \\ y' = c_y \cdot y \end{cases} \tag{8}$$

Calculate the covariance as:

$$\sigma''_{xy} = \frac{1}{n}\sum_{i=1}^{n}(c_x \cdot x_i - c_x \cdot \bar{x})(c_y \cdot y_i - c_y \cdot \bar{y}) = c_x c_y \sigma_{xy} \tag{9}$$

2.2. t-SNE Dimensionality Reduction Algorithm

2.2.1. Introduction to t-SNE Algorithm

t-distributed stochastic neighbor embedding (t-SNE) is a nonlinear machine learning algorithm for dimensionality reduction. It is an improvement of the stochastic neighbor embedding (SNE) [23] proposed by Laurens van der Maaten and Geoffrey Hinton. It is very suitable for high-dimensional data dimensionality reduction to 2D or 3D for visualization. The essence of the process is mapping of the similarity between data points in low-dimensional space to high-dimensional space.

2.2.2. Basic Principles and Derivation of the SNE Algorithm

SNE maps data points to probability distributions by affine transformation. From a mathematical point of view, it can be understood that SNE first converts the Euclidean distance into a conditional probability to express the similarity between points.

Given N high-dimensional data $x_1, x_2, \ldots, x_N$, we first construct a conditional probability p_{ji} proportional to the similarity between x_i and x_j, using the calculation formula Equation (10).

$$p_{ji} = \frac{\exp\left(-\|x_i - x_j\|^2 / \left(2\sigma_i^2\right)\right)}{\sum_{k \neq i} \exp\left(-\|x_i - x_k\|^2 / \left(2\sigma_i^2\right)\right)} \tag{10}$$

In the formula, σ_i is the Gaussian function variance of data point x_i.

For low dimensions y_i and the variance of the Gaussian distribution as $\frac{1}{\sqrt{2}}$, q_{ji} indicates the similarity between two points, as defined in Equation (11):

$$q_{ji} = \frac{\exp\left(-\|x_i - x_j\|^2\right)}{\sum_{k \neq i} \exp\left(-\|x_i - x_k\|^2\right)} \tag{11}$$

As with Equation (10), we assume that $p_{ji} = 0$. If y_i, y_j precisely retain the probability distributions of x_i, x_j, it indicates a better dimension reduction effect, from which Equation (12) is established:

$$p_{ji} = q_{ji} \tag{12}$$

As can be seen from the above, the goal of t-SNE is to find a different way of expressing data that can minimize p_{ij} and q_{ji}. By optimizing the distance between these two probability distributions p_{ij} and q_{ji}, namely KL scatter (Kullback–Leibler divergences), the objective function is given by Equation (13):

$$C = \sum_i KL(P_i \| Q_i) = \sum_i \sum_j p_{ji} \log \frac{p_{ji}}{q_{ji}} \tag{13}$$

In Equation (13), P_i represents the conditional probability distribution of its data points after x_i.

Different points have different σ_i, and the entropy of P_i increases as σ_i increases. SNE uses the concept of perplexity to represent the best σ by binary search. The confusion is:

$$prep(P_i) = 2^{H(P_i)} \tag{14}$$

In Equation (14), $H(P_i)$ is the entropy of P_i:

$$H(P_i) = -\sum_j p_{ji} \log_2 p_{ji} \tag{15}$$

The physical interpretation of confusion is the number of valid neighbors near a point. After determining the value of σ, the problem becomes a solution to the gradient. Therefore, the gradient formula of the objective function of t-SNE can be derived as shown in Equation (16):

$$\frac{\delta C}{\delta y_i} = 4\sum_j \left(p_{ij} - q_{ij}\right)\left(y_i - y_j\right)\left(1 + \|y_i - y_j\|^2\right)^{-1} \tag{16}$$

Generally, the Gaussian distribution under a small σ is used for initialization. In order to speed up the optimization process and avoid falling into the local optimal solution, a large momentum is needed in the gradient update, that is, the parameter update needs to introduce the gradient accumulating term of the previous gradient accumulation. The parameter update formula is Equation (17):

$$Y^{(t)} = Y^{(t-1)} + \eta\frac{\delta C}{\delta Y} + \alpha(t)\left(Y^{(t-1)} - Y^{(t-2)}\right) \tag{17}$$

In Equation (17), $Y^{(t)}$ represents a solution of iteration t times, η represents a learning rate, and $\alpha(t)$ represents a momentum of iteration t times.

2.2.3. t-SNE Principle and Derivation

t-SNE uses the t-distribution in low-dimensional space to characterize the similarity between two points. As can be seen from Figure 2, the red line represents a normal distribution and the blue dashed line represents a t-distribution. Due to the difference in the probability distributions of the normal distribution and the t-distribution, the t-distribution has a longer and longer tail effect than the normal distribution. Therefore, in the high-dimensional space where the data values are relatively compact, the data distribution after the dimensionality reduction can be made larger by using the t-distribution.

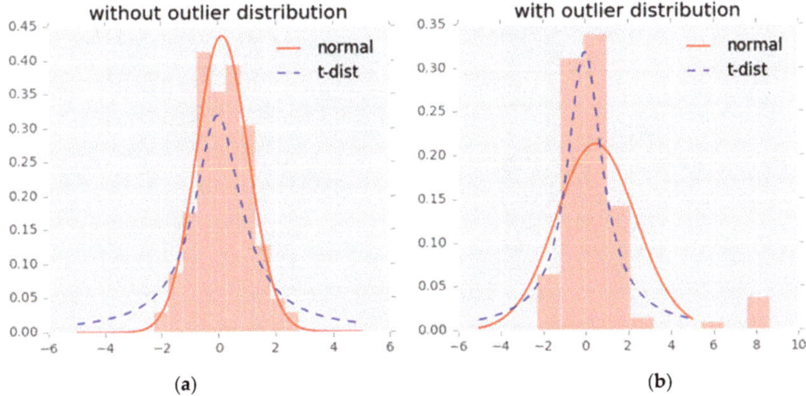

Figure 2. Comparison—normal distribution and t-distribution. (**a**) Distribution without outliers; (**b**) distribution with outliers.

As seen in Figure 2a, in the absence of outliers, the t-distribution can better describe the edge data; as can be seen from Figure 2b, the t-distribution can better reflect the probability distribution of the data in the presence of outliers. The smaller distances are larger than in the normal distribution after mapping, which captures the overall characteristics of the data.

The q_{ij} changes after using the t-distribution can be shown by Equation (18):

$$q_{ij} = \frac{\left(1 + \|y_i - y_j\|^2\right)^{-1}}{\sum\limits_{k \neq l}\left(\left(1 + \|y_i - y_j\|^2\right)^{-1}\right)} \tag{18}$$

In addition, in the computational time complexity, since the t-distribution is a linear sum of Gaussian distributions, it does not increase the time complexity. The post-gradient is optimized by t-distribution, as in Equation (19):

$$\frac{\delta C}{\delta y_i} = 4 \sum_{j} (p_{ij} - q_{ij})(y_i - y_j)(1 + \|y_i - y_j\|^2)^{-1} \tag{19}$$

According to the derivation of the above algorithm, after summarizing, the flow chart of the t-SNE algorithm as shown in Figure 3 can be obtained. The program can then be implemented according to the steps of the algorithm flow chart.

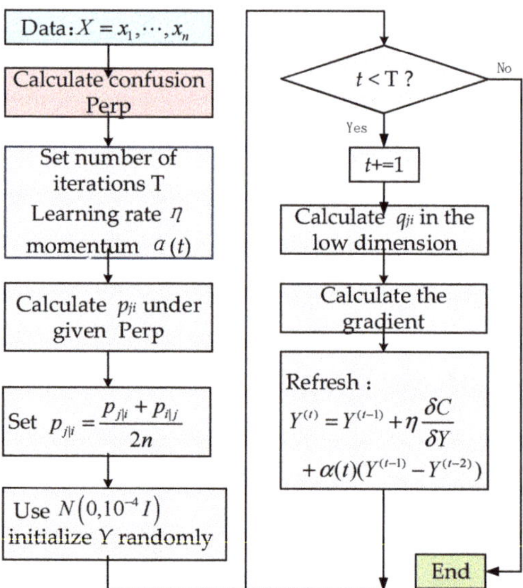

Figure 3. The t-distributed stochastic neighbor embedding (t-SNE) algorithm flow.

3. Weather Forecast Data Preprocessing Scheme and Application

3.1. Composition of Wind Farm Operation Data

Classified by its electrical connection and hardware configuration, wind power big data can be divided into three sources: wind farms, wind turbines, and system access points. The composition of wind power big data is shown in Figure 4. In Figure 4, AGC means automatic generation control, AVC means automatic voltage control, STATCOM means static synchronous compensator.

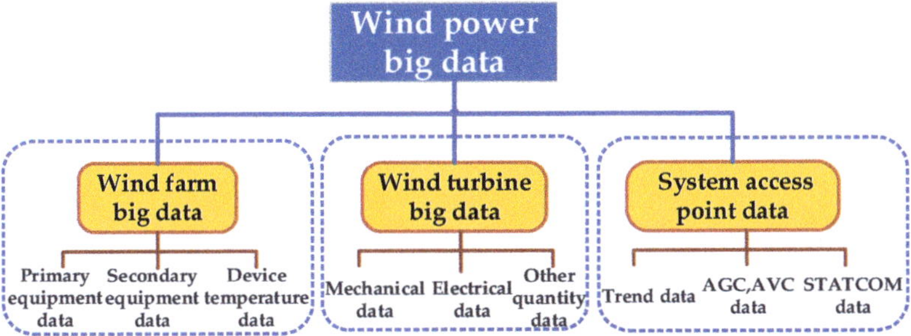

Figure 4. Components in wind power data during operation.

Wind farm big data is composed of primary equipment data, secondary equipment data, and equipment temperature measurement data; wind turbine big data is composed of electrical quantity data, mechanical quantity data, and other data; power system access point big data mainly includes power flow data, AGC, AVC, STATCOM, etc.

The object of this paper is the preprocessing of wind farm NWP data, which is part of the wind turbine big data. As shown in Figure 5, the wind turbine big data mainly includes wind turbine mechanical quantity data, electrical quantity data, and other data including NWP. The mechanical data and electrical data have lower dimensionality and are important operational data, and have no need for dimensionality reduction; NWP data has a high dimensionality which must be considered, so dimension reduction processing must be considered. The dimensionality-reduced data can be applied to the analysis of various problems, including forecasting, running scheduling, and so on.

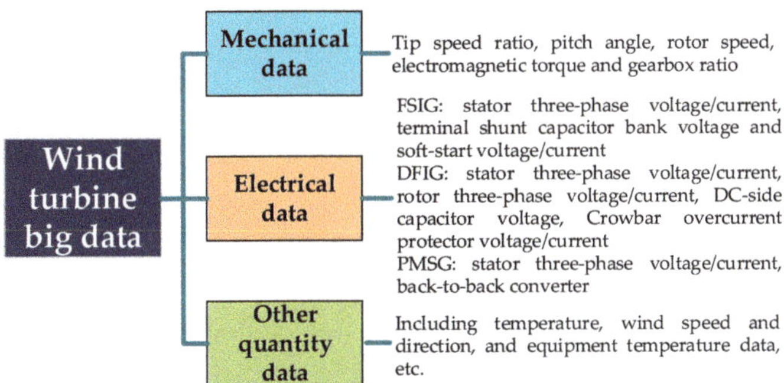

Figure 5. The data components from wind turbines.

For the prediction of wind power output, the current research focused on the wind speed–power model. However, considering only the influence of wind speed on power, other indicators related to power output may be ignored, resulting in a decrease in prediction accuracy. Detailed NWP data such as wind direction, wind speed, atmospheric pressure, and air density of wind farms are used as references for dimension reduction processing, which plays an important role in wind power forecasting and operation scheduling.

3.2. Numerical Weather Data Acquisition and Processing Steps

Numerical weather forecast data has a large number of meteorological indicator variables. The processing method used to date is to select meteorological indicator variables according to experience, but the accuracy of selecting meteorological indicators by experience alone cannot be guaranteed. In addition, low correlation or redundant variables will also adversely affect the cost and time of prediction. In order to improve the efficiency of the model, the N-dimensional data were reduced by the t-SNE dimensionality reduction method.

The purpose of preprocessing wind power data is to normalize, reduce dimensionality, and predict the error of the collected NWP data. The specific implementation steps are as follows:

(1) The operating status of N fans is measured by various sensors installed on the fan and uploaded to the main station of the wind airport.

(2) The dimension equivalent of each dimension of the collected data is calculated according to Equations (1)–(10) to avoid the influence of different dimensions.

(3) According to Equations (11)–(19), the dimensionality of the different indicators in the NWP data is reduced to reduce the redundancy of the phase data.

(4) The effectiveness of the preprocessing is verified by using the LSTM network for power prediction.

(5) Visualize the forecast data and historical data.

4. Case Analysis

4.1. Data Source

The sample data used in this paper were from the data segment collected by a wind turbine. The sampling start time was 13:33 on 6 August 2013, and a total of 2.4 million pieces of data were collected. After eliminating the missing variables, the NWP data has 22 remaining dynamics, pressure, temperature, humidity, wind speed, and wind direction at different heights, as shown in Table 1.

Table 1. Numerical weather prediction NWP variables.

Number of Variables	Index	Shorthand	Unit
1	Air pressure	p	mbar
2	Air temperature	T	degC
3	Thermodynamic temperature	$Tpot$	K
4	Relative humidity	rh	%
5	Relative pressure	$VPact$	mbar
6	Absolute humidity	sh	g/kg
7	Air density	rho	g/m^3
8	Minimum wind speed	$minWs$	m/s
9	Rainfall	TP	mm
10	Photosynthetically active radiation	PAR	μmol/m/s
11	Logarithmic temperature	$Tlog$	degC
12	Carbon dioxide concentration	$CO2$	ppm
13	Maximum wind speed	$maxWs$	m/s
14	10 m wind speed	wv	m/s
15	10 m wind direction	wd	deg
16	10 m wind level	ws	/
17	20 m wind speed	wv	m/s
18	20 m wind direction	wd	deg
19	20 m wind rating	ws	/
20	30 m wind speed	wv	m/s
21	30 m wind direction	wd	deg
22	30 m wind rating	ws	/

As can be seen from Table 1, the dimensionality of the NWP data was very high, and it was not possible to determine whether each feature affects the operating state of the wind farm. If the data were subsequently input into the prediction model without processing, too many data would not only lead to a large increase in computation time, but would also affect the ability of the model to express features. Therefore, it was necessary to select valuable features from the appropriate algorithms. In this part, the t-SNE method introduced above was used for feature selection, and compared with the PCA dimensionality reduction algorithm.

4.2. Wind Power Big Data Dimensionality Reduction Based on t-SNE Algorithm

In order to remove the noise of the NWP samples and visually reflect the characteristics of wind farm meteorological data in low-dimensional space, the sample set was reduced from 22 dimensional to 2 dimensional space using the t-SNE algorithm, the confusion was set to 20, and iteration was set to 5000 times. The effect of confusion was to balance the weights of the t-SNE local transformation and the global transformation. It can be understood that the confusion was used to set the number of adjacent points of each point. The greater the confusion setting, the more attention is paid to the global data distribution. Usually, the confusion parameter is roughly equal to the number of neighbors needed. In this paper, it was determined based on the NWP variables. The number of iterations was based on the parameters recommended by the authors in the literature [23].

We selected 3000 data from a single day to show the dimensionality reduction results of the t-SNE algorithm, as shown in Figure 6. The dots in the figure represent data points, and the different colors represent different variables.

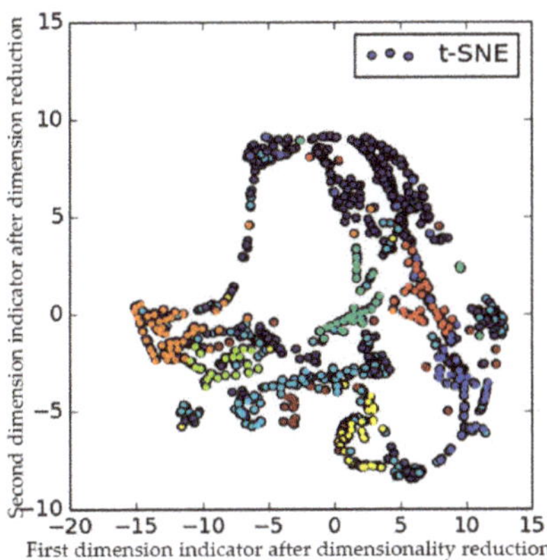

Figure 6. The 2 dimensional result of t-SNE dimensionality reduction algorithm.

As shown in Figure 6, the data of the input NWP were color-coded according to the number of data categories in the default series of RGBA, RGBA is the color space representing red green blue and alpha. The t-SNE algorithm was able to clearly represent all data points in a 2 dimensional space, and most of the data points of different features exhibited a short-line structure of one or several segments. The t-SNE algorithm clearly separated the different categories of data.

At the same time, it can be seen from Figure 6 that when the algorithm was used to reduce the original data to 2 dimensions, some data points overlapped, for example, the red and blue in the figure overlap, making them more difficult to distinguish. Therefore, the following attempt was to to reduce the original sample set to the three dimensional subspace using the t-SNE algorithm. The confusion was set to 20 and iteration was set to 5000 times. We again selected 3000 data to show the dimensionality reduction results, as shown in Figure 7. The colors of the input data were color-coded according to the format of "RdYlGn", which is the order from red to green.

In order to verify the generalization of the t-SNE model, the meteorological data segment of this wind farm at other times was used as the experimental object, and the sampling start time was 8 August 2013. After the data were input into the model using the same preprocessing method, the dimensionality reduction visualization that resulted is shown in Figure 8.

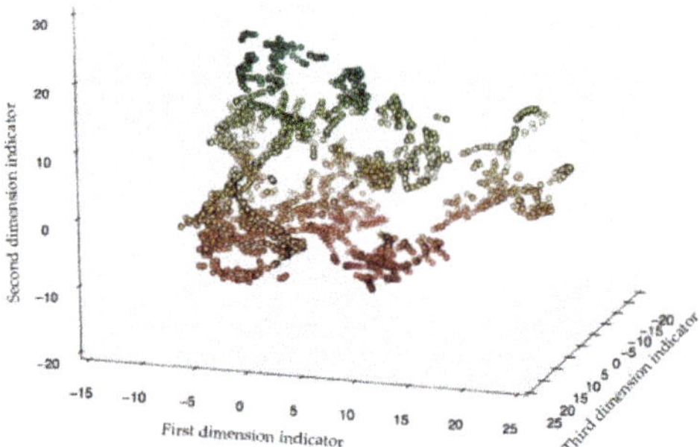

Figure 7. The three dimensional result of t-SNE dimensionality reduction algorithm.

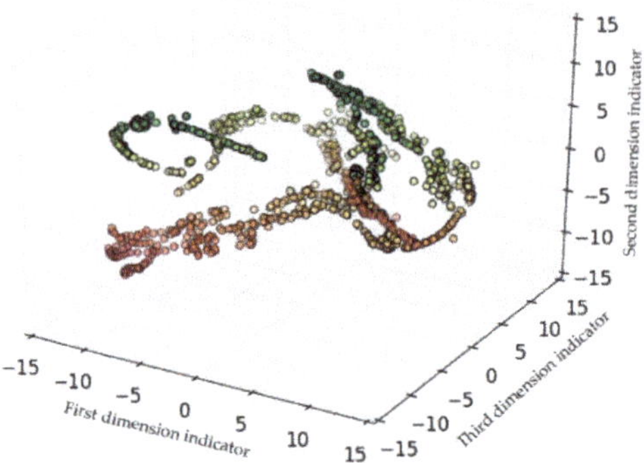

Figure 8. The three dimensional result of t-SNE dimensionality reduction algorithm for data series 2.

It can be seen from the above simulation results that the t-SNE algorithm could clearly represent all data points in three dimensional space. Most data points presented a one- or several-segment short-line segment structure that reflected the temporal continuity of weather changes. It can be seen that dots of different colors represent different features that were distinctly distinguished in three dimensions. The simulation results demonstrated the effectiveness of the t-SNE algorithm in processing meteorological data in wind farm operating data.

Analysis of the distance relationship between data points does not provide quantitative information about the data. Therefore, the purpose of the t-SNE dimensionality reduction method is mainly to visualize the data, so that we can have a macroscopic understanding of the data patterns that need to be mined. For a certain set of data, if t-SNE performs well on the segmentation feature, it is highly probable that a machine learning method that projects this set of data into different categories will be found. Conversely, if t-SNE is generally represented on segmentation features, such as in the case of class overlaps, then a more complex model needs to be built.

4.3. Comparison with the PCA Algorithm for Dimensionality Reduction

The idea of principal component analysis is to find one or several projection directions so that the variance of the original data samples after projection is maximized. The original m-dimensional features are projected onto a new n-dimensional space, which is characterized by the principal component. The main evaluation method of principal component selection is to use variance. The larger the variance of new features, the more information contained in this feature can be reflected. Therefore, the percentage of contribution of cumulative variance is calculated to select the principal component.

Assuming that the sample set $X = \{x_1, x_2, \ldots, x_m\}$ satisfies the centralization, it is assumed that the new coordinate system after the projection transformation is $\{w_1, w_2, \ldots, w_d\}$, where w_i is the standard orthogonal basis vector and $\|w_i\|_2 = 1$. The projection of a data point x_i in the new coordinate system $\{w_1, w_2, \ldots, w_d\}$ is $W^T x_i$. If the projection of the data points in the original sample set can be effectively separated under this new coordinate system, the variance of the different sample data points in the new coordinate system is $\sum_i W^T x_i x_i^T W$, so the optimization goal is to maximize this variance:

$$\begin{cases} \max_w \mathrm{tr}\left(W^T X X^T W\right) \\ \text{s.t.} \quad W^T W = 1 \end{cases} \tag{20}$$

For Equation (20), the Lagrangian multiplier method is used, giving:

$$X X^T W = \lambda W \tag{21}$$

Therefore, it is only necessary to perform eigenvalue decomposition on the covariance matrix and sort the obtained eigenvalues: $\lambda_1 \geq \lambda_2 \geq \ldots \geq \lambda_m$. Generally, a dimension with a cumulative contribution rate of about 75% to 95% is selected as the reference dimension after PCA dimensionality reduction.

The variance contribution rate and the cumulative variance contribution rate are, respectively:

$$\eta_i = \frac{100\% \lambda_i}{\sum_m \lambda_i} \tag{22}$$

$$\eta_\Sigma(p) = \sum_i^p \eta_i \tag{23}$$

The eigenvectors corresponding to the first x eigenvalues constitute the solution of principal component analysis $W = (w_1, w_2, \ldots, w_x)$.

In order to compare the dimensionality reduction effect of the t-SNE and PCA algorithms, the data of wind farm meteorological data segment 1 were reduced to 2D, 3D, 5D, and 8D space, and credibility was used as the evaluation standard. Credibility indicates the retention of the local structure of the original structure of the data when dimension reduction to low-dimensional space is carried out. The size range of credibility is [0,1]. The greater the credibility, the better the data retention, and the lower the credibility, the worse the data retention after dimension reduction. The mathematical definition of credibility is given by Equation (24).

$$T(k) = 1 - \frac{2}{nk(2n - 3k - 1)} \sum_{i=1}^{n} \sum_{j \in u_i} (r(i, j) - k) \tag{24}$$

In Equation (24), $r(i, j)$ represents the rank of the low-dimensional data points j, determined according to the pairwise distance between the low-dimensional data points, and U_i^k represents the set of neighbor data points k in the low-dimensional space. The following will be used to compare the reliability of high-dimensional data to 2, 3, 5, and 8 dimensions using PCA and t-SNE.

Table 2 and Figure 9 show the comparison of the reliability of the data after dimension reduction using the t-SNE algorithm and the PCA method. Through the graph, it can be seen that t-SNE gave a significant improvement in the dimensionality reliability of the experimental low-dimensional space compared with PCA, and t-SNE basically retained the time-series characteristics of the original data. PCA means principal component analysis

Table 2. Comparison of trustworthiness of low-dimensional representations of the data set. PCA—principal component analysis.

Dimension	PCA	t-SNE
2	0.918	0.921
3	0.970	0.975
5	0.975	0.983
8	0.977	0.989

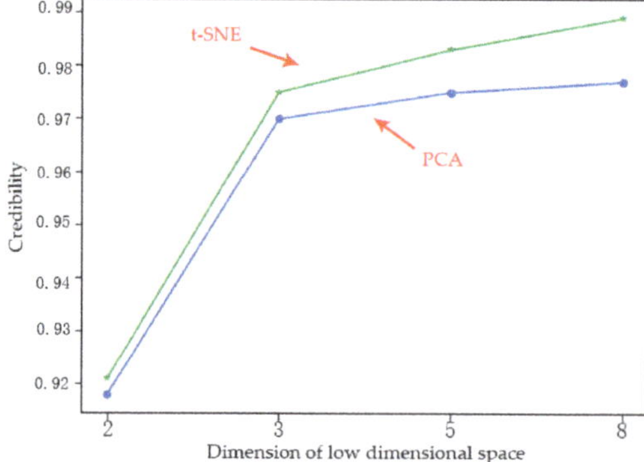

Figure 9. Dimensionality reduction results of principal component analysis PCA and t-SNE.

4.4. Comparison of Wind Speed Prediction Before and After Data Preprocessing

Here, the long-short-term memory (LSTM) was selected as the wind speed prediction model to evaluate the effect of the wind power data preprocessing. As a complex nonlinear unit, LSTM uses a deeper neural network to reflect long-term memory effects and has deep learning ability [24,25].

The preprocessed data were divided into training data and test data. Among them, 1300 pieces of data are used as training data, and the remaining 500 pieces of data are used as test data.

In the error analysis of the prediction results, it is often evaluated by two evaluation indicators: mean absolute percentage error (MAPE) and root mean square error (RMSE). The error calculation formula is given by reference to Equations (25) and (26), respectively.

$$\varepsilon_{MAPE} = \frac{1}{n}\sum_{i=1}^{n} \frac{\left|\hat{P}_N(i) - P_N(i)\right|}{P_N(i)} \times 100\% \tag{25}$$

$$\varepsilon_{RMSE} = \sqrt{\frac{1}{n}\sum_{i=1}^{n} \left(\hat{P}_N(i) - P_N(i)\right)^2} \tag{26}$$

In Equations (25) and (26), $P_N(i)$ and $\hat{P}_N(i)$ ($i = 1, 2, 3, \ldots, n$) are the actual measured and predicted values of the data point i, respectively, and n represents the length of the data used for verification.

Table 3 shows the prediction results of the wind farm data through the preprocessing method of this paper and the direct use of the original data. It can be seen from Table 3 that the prediction results ε_{MAPE} and ε_{RMSE} after preprocessing by t-SNE were reduced compared with the prediction results using historical data, which effectively improved the prediction accuracy. The results also show that after the dimension reduction preprocessing, the analysis of less relevant invalid variables can be avoided, and only the highly correlated useful variables were retained, which helps to improve the prediction performance of the LSTM model. In addition, after using the dimensionality reduction preprocessing method of this paper, the input variables were much fewer than the original, which is conducive to large-scale data calculation. MAPE is mean absolute percentage error and RMSE means root mean square error.

Table 3. Error analysis of the forecasting result.

Index	Type of Data	ε_{MAPE} (%)	ε_{RMSE}
Active power	Preprocessed data	0.603	2098.866
	historical data	0.711	2293.650
Phase current	Preprocessed data	3.1589	73.358
	historical data	3.722	80.577
Phase voltage	Preprocessed data	2.224	32.68
	historical data	2.517	37.955

4.5. Visualization Platform Implementation

We designed a visualization system for statistical and real-time status monitoring of wind power big data. In order to display relevant information in a timely manner, the platform uses Grafana as a visualization tool and the timing database InfluxDB as a data storage container. In the experimental part, the Python language was used to implement various functions, including client and server building, reading, and writing to InfluxDB.

InfluxDB is backed by Norwest Venture Partners, Sapphire Ventures, Battery Ventures, Trinity Ventures, Mayfield, Harmony Partners, Sorenson Capital, Bloomberg Beta and Y Combinator, its location is San Francisco, CA 94103, USA. Grafana is created by raintank co-founder Torkel Odegaard and located in San Francisco, USA. Python is created by Guido van Rossum and managed by Python software foundation, located in Beaverton 97008, USA.

4.5.1. System Architecture and Implementation Process

The overall architecture of the wind farm monitoring data visualization platform is shown in Figure 10. The visualization platform was mainly composed of a data processing module and a data visualization module. The data processing module was responsible for processing the raw data and importing it into the database. The visualization module was responsible for reading and aggregating the data and visualizing it. The data visualization module also included data query and data aggregation functions. Through these two functions, the wind farm monitoring data visualization platform can be realized.

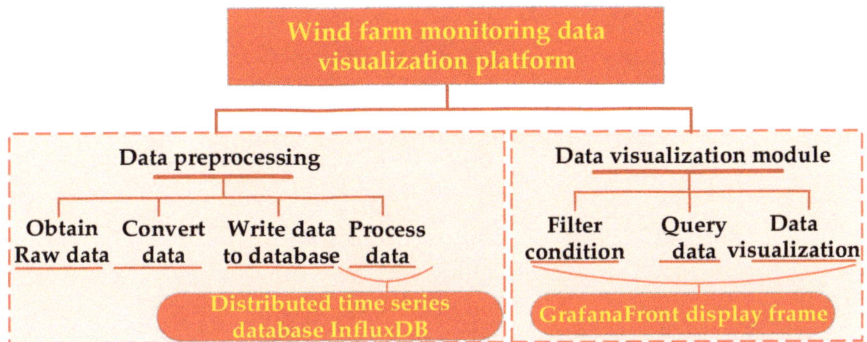

Figure 10. Overall framework of the system.

The visualization implementation process is shown in Figure 11. The data were processed and filtered, transformed into visually expressible geometric data by mapping, and finally rendered into user-visible image data.

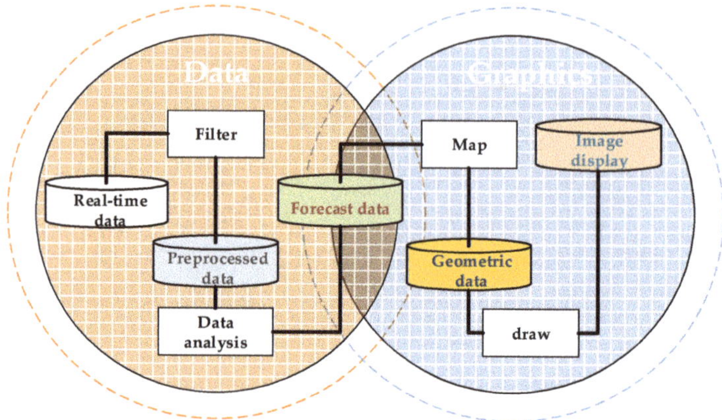

Figure 11. Mapping from data space to graphics space.

4.5.2. Visualization Platform Implementation

The data visualization platform included the following three modules: a data processing module, a data aggregation module, and a data visualization module. The data processing module converted the wind power data in the form of a csv file into Json format and wrote it to the InfluxDB database in batches. The data aggregation module compressed aggregated operational data through the data retention function and continuous query (CQ) function provided by InfluexDB. Data visualization module: Connect the data in the InfluxDB database to Grafana and select the appropriate visualization panel to visualize meteorological data such as precipitation, pressure, temperature, humidity, and wind speed and direction.

Currently, there are six types of panels, including Graph, Singlestat, Heatmap, Dashlist, Table, and Text. The visualization panel for each meteorological factor of the wind farm is shown in Figure 12.

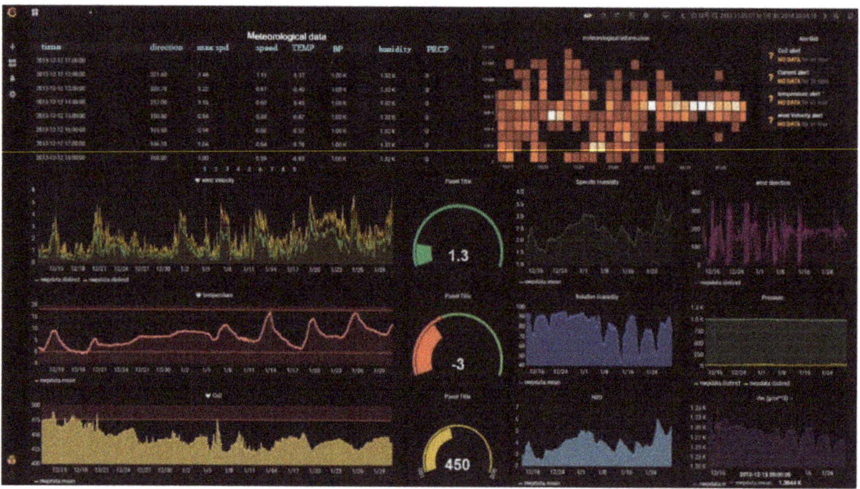

Figure 12. Panel of numerical weather prediction NWP data of wind farm.

5. Conclusions

In this paper, the preprocessing links in wind farm big data mining were studied, and data preprocessing methods were discussed and applied. The t-SNE algorithm was used to preprocess and analyze numerical weather prediction (NWP) data. The main conclusions are as follows:

(1) Due to the large size of meteorological indicator variables in NWP data, the traditional feature selection method is no longer effective. For this reason, the t-SNE algorithm was used to reduce the NWP data. Using actual NWP data collected by a wind farm, the experiment proved that t-SNE can better preserve the local similarity of sample points in the original high-dimensional space in 2 dimensional space; the t-SNE data preprocessing method improved the computational efficiency of the subsequent data analysis model while ensuring accuracy.

(2) By comparing two different data preprocessing methods, t-SNE and PCA, it was found that the dimensionality reliability of t-SNE was slightly better than the PCA dimensionality reduction method in each low-dimensional space of the experiment; the data preprocessing results of the t-SNE and PCA algorithms were applied to wind power prediction based on a deep learning LSTM network, which proved that the preprocessed data had better prediction accuracy.

(3) The wind farm monitoring data visualization platform consisted of a data processing module, a data aggregation module, and a data visualization module, able to realize the visualization of the massive data recorded during the operation of the wind farm. It not only provides important understanding of the operating state of the wind farm, but also provides a basis for the construction of subsequent trend prediction models.

Author Contributions: Conceptualization, D.X. and Y.W.; Methodology, J.G.; Validation, Y.W., J.G. and Y.Z.; Formal Analysis, J.G.; Investigation, Y.W.; Data Curation, Y.Z.; Writing—Original Draft Preparation, J.G.; Writing—Review & Editing, J.G.; Project Administration, D.X.; Funding Acquisition, Y.Z.

Funding: This research was supported by the National Natural Science Foundation of China (51677114). State Grid Project: Study on the Mechanism and Suppression Measures of Complex Oscillation in Multi-source Scenery of Wind Power, Photovoltaic and Thermal Power (SGTYHT/16-JS-198).

Conflicts of Interest: The authors declare no conflict of interest.

References

1. Xia, F.; Song, F. Evaluating the economic impact of wind power development on local economies in China. *Energy Policy* **2017**, *110*, 263–270. [CrossRef]

2. Lyu, B.; Li, Y.; Fu, J.; Trapp, A.C.; Xie, H.; Liao, Y. Scalable User-Substation Assignment with Big Data from Power Grids. *IEEE Trans. Big Data* **2017**, *5*, 209–222. [CrossRef]

3. Wang, Z.; Wang, W.; Liu, C.; Wang, B.; Feng, S. Probabilistic forecast for aggregated wind power outputs based on regional NWP data. *J. Eng.* **2017**, *2017*, 1528–1532. [CrossRef]

4. Prusty, B.R.; Jena, D. Preprocessing of Multi-Time Instant PV Generation Data. *IEEE Trans. Power Syst.* **2018**, *33*, 3189–3191. [CrossRef]

5. Zheng, L.; Hu, W.; Min, Y. Raw Wind Data Preprocessing: A Data-Mining Approach. *IEEE Trans. Sustain. Energy* **2014**, *6*, 11–19. [CrossRef]

6. Zhang, Y.; Zhao, Y.; Gao, S. A Novel Hybrid Model for Wind Speed Prediction Based on VMD and Neural Network Considering Atmospheric Uncertainties. *IEEE Access* **2019**, *7*, 60322–60332. [CrossRef]

7. Wan, C.; Xu, Z.; Pinson, P.; Dong, Z.Y.; Wong, K.P. Optimal Prediction Intervals of Wind Power Generation. *IEEE Trans. Power Syst.* **2013**, *29*, 1166–1174. [CrossRef]

8. Lee, D.; Baldick, R. Short-Term Wind Power Ensemble Prediction Based on Gaussian Processes and Neural Networks. *IEEE Trans. Smart Grid* **2013**, *5*, 501–510. [CrossRef]

9. Yan, J.; Li, K.; Bai, E.-W.; Deng, J.; Foley, A.M. Hybrid Probabilistic Wind Power Forecasting Using Temporally Local Gaussian Process. *IEEE Trans. Sustain. Energy* **2015**, *7*, 87–95. [CrossRef]

10. Tanasa, D.; Trousse, B. Data preprocessing for wum. *IEEE Potentials* **2004**, *23*, 22–25. [CrossRef]

11. Wang, Z.; Wang, C.; Wu, J. Wind energy potential assessment and forecasting research based on the data pre-processing technique and swarm intelligent optimization algorithms. *Sustainability* **2016**, *8*, 1191. [CrossRef]
12. Niu, X.; Wang, J. A combined model based on data preprocessing strategy and multi-objective optimization algorithm for short-term wind speed forecasting. *Appl. Energy* **2019**, *241*, 519–539. [CrossRef]
13. Jiang, P.; Ma, X. A hybrid forecasting approach applied in the electrical power system based on data preprocessing, optimization and artificial intelligence algorithms. *Appl. Math. Model.* **2016**, *40*, 10631–10649. [CrossRef]
14. Tian, C.; Hao, Y.; Hu, J. A novel wind speed forecasting system based on hybrid data preprocessing and multi-objective optimization. *Appl. Energy* **2018**, *231*, 301–319. [CrossRef]
15. Malvoni, M.; De Giorgi, M.G.; Congedo, P.M. Forecasting of PV Power Generation using weather input data-preprocessing techniques. *Energy Procedia* **2017**, *126*, 651–658. [CrossRef]
16. Azimi, R.; Ghofrani, M.; Ghayekhloo, M. A hybrid wind power forecasting model based on data mining and wavelets analysis. *Energy Convers. Manag.* **2016**, *127*, 208–225. [CrossRef]
17. Zhao, Y.; Ye, L.; Wang, W.; Sun, H.; Ju, Y.; Tang, Y. Data-Driven Correction Approach to Refine Power Curve of Wind Farm Under Wind Curtailment. *IEEE Trans. Sustain. Energy* **2017**, *9*, 95–105. [CrossRef]
18. Ye, X.; Lu, Z.; Qiao, Y.; Min, Y.; O'Malley, M. Identification and Correction of Outliers in Wind Farm Time Series Power Data. *IEEE Trans. Power Syst.* **2016**, *31*, 4197–4205. [CrossRef]
19. Renani, E.T.; Elias, M.F.M.; Rahim, N.A. Using data-driven approach for wind power prediction: A comparative study. *Energy Convers. Manag.* **2016**, *118*, 193–203. [CrossRef]
20. Ullah, N.; Zameer, A.; Khan, A.; Javed, S.G. Machine Learning based short term wind power prediction using a hybrid learning model. *Comput. Electr. Eng.* **2015**, *45*, 122–133.
21. Yuan, X.; Chen, C.; Yuan, Y.; Huang, Y.; Tan, Q. Short-term wind power prediction based on LSSVM–GSA model. *Energy Convers. Manag.* **2015**, *101*, 393–401. [CrossRef]
22. Abdoos, A.A. A new intelligent method based on combination of VMD and ELM for short term wind power forecasting. *Neurocomputing* **2016**, *203*, 111–120. [CrossRef]
23. Maaten, L.; Hinton, G. Visualizing data using t-SNE. *J. Mach. Learn. Res.* **2008**, *85*, 2579–2605.
24. Barbounis, T.; Theocharis, J.; Alexiadis, M.; Dokopoulos, P. Long-Term Wind Speed and Power Forecasting Using Local Recurrent Neural Network Models. *IEEE Trans. Energy Convers.* **2006**, *21*, 273–284. [CrossRef]
25. Mandal, P.; Srivastava, A.K.; Park, J.-W. An Effort to Optimize Similar Days Parameters for ANN-Based Electricity Price Forecasting. *IEEE Trans. Ind. Appl.* **2009**, *45*, 1888–1896. [CrossRef]

Article

A Decentralized Architecture Based on Cooperative Dynamic Agents for Online Voltage Regulation in Smart Grids

Amedeo Andreotti [1], **Alberto Petrillo** [1,*], **Stefania Santini** [1] **and Alfredo Vaccaro** [2] **and Domenico Villacci** [2]

[1] Department of Information Technology and Electrical Engineering (DIETI), University of Naples Federico II, 80125 Naples, Italy; amedeo.andreotti@unina.it (A.A.); stefania.santini@unina.it (S.S.)

[2] Department of Engineering, University of Sannio, 82100 Benevento, Italy; vaccaro@unisannio.it (A.V.); villacci@unisannio.it (D.V.)

* Correspondence: alberto.petrillo@unina.it

Received: 1 March 2019; Accepted: 2 April 2019; Published: 10 April 2019

Abstract: The large-scale integration of renewable power generators in power grids may cause complex technical issues, which could hinder their hosting capacity. In this context, the mitigation of the grid voltage fluctuations represents one of the main issues to address. Although different control paradigms, based on both local and global computing, could be deployed for online voltage regulation in active power networks, the identification of the most effective approach, which is influenced by the available computing resources, and the required control performance, is still an open problem. To face this issue, in this paper, the mathematical backbone, the expected performance, and the architectural requirements of a novel decentralized control paradigm based on dynamic agents are analyzed. Detailed simulation results obtained in a realistic case study are presented and discussed to prove the effectiveness and the robustness of the proposed method.

Keywords: voltage regulation; smart grid; decentralized control architecture; multi-agent systems

---

1. Introduction

The intermittent power profiles generated by renewable power generators in power grids considerably perturb the bus voltage magnitude [1–3], which may go outside the allowable admissible range, especially during critical operating conditions (i.e., high generation and low load demand [3–5]). These events are not infrequent in existing power networks, which have been traditionally designed by assuming the passivity of all system buses, without considering the presence of distributed and dispersed generators [6]. Hence, increasing of renewable power generators in these networks, driven by the modern sustainable environmental policies, is causing severe and complex phenomena that need to be carefully addressed [7–9]. In this context, the research for new and more advanced online voltage control systems, aimed at regulating the reactive power injected/absorbed by distributed generators, represents one of the most promising research directions for smart power grids [10–12]. To address this issue, the use of centralized architectures, traditionally used for voltage control in power systems, could not be suitable in short-term scenarios [13], since it asks for a significant upgrade of the communication and computing resources, to effectively solve constrained optimal power flow problems [14]. Moreover, the deployment of reliable state estimation algorithms, which is a prerequisite of optimal power flow-based regulation techniques, is still an open problem due to the limited number of installed sensors [15]. This has stimulated the research for alternative control techniques for the coordination of the reactive power injected/absorbed by distributed generation units, which process only local data [16–18]. Recent experimental results have demonstrated that these local control techniques could be effectively used in existing distribution networks, but they may not provide sufficiently accurate results, especially in the presence of many renewable power generators [11]. To solve this problem, in [19–21] the voltage regulation problem is solved by a centralized computing framework, which collects and processes the sensor data streaming, and sends back the corresponding set-points to the voltage controllers. Although the results obtained by this hierarchical voltage control paradigm are highly accurate, a reliable and pervasive communication system is required to connect all the sensor/controllers to the central processing system, which makes the entire control architecture extremely vulnerable. Furthermore, this method asks for an accurate model of the power system and a reliable estimation of the load/generation patterns, which are highly unpredictable, due to rapidly changing operating conditions [22]. Moreover, as recently outlined in many research works, these kinds of centralized control architectures are characterized by low scalability levels, since an increased grid complexity could ask for unaffordable computing resources, further hardware redundancies, higher communication bandwidth, and larger data storage resources [23]. All these limitations could hinder the application of centralized control architectures in modern power systems, where the constant growth of grid complexity and the need for a massive pervasion of renewable power generators ask for more scalable, and more flexible control architectures. In this framework, the deployment of decentralized architectures based on cooperative controllers, which infer global information about the actual power system operation by exchanging and processing only local data, has been recognized as one the most promising enabling technology for solving the voltage control problem. The concept is to try keeping the bus voltage magnitudes very close to the nominal value by adjusting the reactive power generated by the distributed generators, without asking for sophisticated communication hardware and computational resources. In particular, in [24,25] a decentralized technique for voltage regulation based on mutually coupled oscillators has been proposed. The main idea is to couple each grid sensor to a first-order oscillator equipped with decentralized consensus protocols, which converge to the global variables characterizing the actual power system operation. Thanks to this paradigm, the distributed voltage controllers can infer the global variables without the need for a fusion center, which collects and processes all the sensor data. In [26], a two-stage control technique for decentralized voltage regulation in active networks has been proposed. During the first stage, all the local controllers adjust the voltage at the monitored bus, by only processing local data; in the second stage, a proper coordination strategy is activated to properly balance the reactive power absorbed/injected by each controller. A similar approach is proposed in [19], where two control techniques are conceived, to avoid excessive reactive power absorption/injection by each distributed controller. In particular, the first technique is based on the decentralized estimation of the average reactive power generated by all the controllers. The second approach is based on the estimation of the average current that controllers inject at the point of common coupling, which is then used to define the optimal set of each controller. More recently, new techniques based on multi-agent systems (MASs) have been proposed for decentralized voltage regulation. In particular, in [27] a MAS-based decentralized technique for voltage control in distribution networks, and an incentive mechanism aimed at stimulating renewable power generators to support the grid voltage are designed. The main idea is to allow the local control agents to compute the voltage sensitivities by cooperating only with their neighborhood, without the need for an arbitration agent, which collect all the agent's measurements. Starting from these sensitivity coefficients, each agent identifies its local voltage control strategy by maximizing its own profit. An alternative decentralized approach to compute the voltage sensitivities, which is based on the data surface fitting technique, has been proposed in [28]. This technique requires the knowledge of the network characteristics, and it is robust to changes in network parameters. The decentralized optimization of the local voltage control strategy is obtained in [29] by employing a computing paradigm, which aims at minimizing a quadratic voltage mismatch error objective using gradient-projection updates. To solve this problem two dynamic scenarios have been considered, which include an asynchronous scheme for the decentralized parameters update, and a time-varying communication scheme for highly variable network operation states. A similar solution approach has been proposed in [30] for voltage control of radial distribution grids with photovoltaic generators operating in voltage support mode, and distributed storage systems. In this paper, the voltage optimization problem has been led to the design of a decentralized disturbance-feedback controller, which minimizes the expected value of a convex quadratic cost function, subject to robust convex quadratic constraints on the system state and input. This decentralized problem has been solved by deriving an inner approximation, which enables the efficient computation of an affine control policy via the solution of a finite-dimensional conic program. Although these techniques show their potential of decentralized architectures in solving the online voltage regulation, their performance, in terms of grid voltage magnitude deviations, may be not satisfactory [31]. This limitation mainly stems from the fact that the computed control action is not based on a real picture of the system operating condition. Hence, new and more effective decentralized voltage control architectures should be looked for. In trying to address these problems, this paper proposes a novel decentralized control architecture for the online voltage regulation in active power grids. The proposed solution leverages the mathematical framework of consensus/synchronization control theory in MASs. MASs consist of groups (i.e., ensemble) of dynamical systems exchanging their information and interacting with each other through wireless/wired communication networks, to agree, for example, upon a certain quantity of interest. Many real systems in nature and human society can be modeled as MAS and many researchers, inspired by natural occurrence of flocking and formation forming, have focused their work on synchronization, consensus and coordination of these latter [32], i.e., in controlling the whole network in order to produce a common behavior by applying distributed algorithms and to guarantee a smart group behavior. Examples in engineering deal with the coordinated motion of autonomous vehicles [33–35], the phase or frequency synchronization in large power grids [36], and the synchronization of wireless sensor networks [37].

By leveraging this paradigm, an electrical grid can be controlled by exploiting N cooperative smart agents that, exchanging their state information through communication networks, impose a common voltage magnitude value to the whole electrical grid. Specifically, each smart controller uses the information received by neighbors, via a communication interface, to locally control the voltage magnitude of the monitored bus, while aiming at achieving a synchronization behavior to desired voltage magnitude, as imposed by the network generator without the need for fusion data center. This allows the smart controller to locally decide how much reactive power it needs to inject, or absorb, to reach the desired voltage asset for the whole power grid.

The current literature exploits decentralized control strategies only for the control of the generators (see the survey [31] and the references herein). Our approach, differently, aims to guarantee that all the buses, both generation and load, synchronize to the common reference imposed by the generators, while reducing power losses. The proposed approach hence provides a self-organized power grid, which has the ability to cope effectively with the problems that might occur using only local interactions and providing "plug and play" capability. A case study developed on the IEEE 30-bus network confirm the effectiveness of the approach, and shows its robustness regarding electrical load variations.

Finally, it is worth mentioning that the large-scale deployment of the proposed control paradigm asks for the conceptualization of new tools aimed at facilitating operational data acquisition and handling in inter-operable formats. In this domain the current research activities are oriented in conceptualizing advanced computing paradigms aimed at handling complex systems by information semantics [38]. The main idea is to develop a power system ontology able to adapt to different use-cases, which could be used to define and specify complex events and actions that run on an event processing engine. To this aim the Authors are developing a distributed and cooperative information framework aimed at exploiting the semantic representation of power system measurements for transparently exchanging data and information between the local voltage controllers, and with the Energy Management Systems of the Distributed System Operator. The proposed framework is based on specific software components, which support decentralized semantic sensor data exchanging and a distributed middleware aimed at processing massive and heterogeneous sensor measurements. The details of this framework will be presented in a future works.

The paper is organized as follows. In Section 2 the problem is formulated, while in Section 3 the proposed decentralized solution is analyzed. Section 4 is devoted to a case study; concluding remarks are presented in Section 5.

2. Problem Formulation

The online voltage control function identifies, for each power system state Γ, the set-point y of the grid controllers that minimizes an objective function J subject to several equality and inequality constraints $g(\Gamma, y)$.

This problem can be formalized by the following constrained optimization problem:

$$\left\{ \begin{array}{l} \min_{y \in \Omega} J(y, \Gamma) \\ g(\Gamma, y) \leq 0 \end{array} \right. \tag{1}$$

where

$$y = \left[Q_{dg,1}, \cdots, Q_{dg,N_g}, Q_{cap,1}, \cdots, Q_{cap,N_c}, m \right] \tag{2}$$

is the control vector, whose components are the grid controller set-points, $Q_{dg,i}$ is the reactive power injected by the i-th ($i = 1, \ldots, N_g$) distributed generator available for the regulation; $Q_{cap,j}$ is the vector of the reactive power injected by the j-th ($j = 1, \ldots, N_c$) capacitor bank and m is the tap position of the HV/MV line tap changing transformer. Please note that the control vector y takes value in the solution space Ω:

$$y \in \Omega \Longleftrightarrow \left\{ \begin{array}{l} tap_{min} \leq m \leq tap_{max}, \\ Q_{dg,min,i} \leq Q_{dg,1} \leq Q_{dg,max,i} \quad i = 1, \cdots, N_g, \\ Q_{cap,min,j} \leq Q_{cap,j} \leq Q_{cap,max,j} \quad j = 1, \ldots, N_c. \end{array} \right. \tag{3}$$

The vector function $g(\Gamma, y)$ describes the problem constraints in terms of allowable ranges for the bus voltage magnitudes (i.e., $V_{min,q} \leq V_q \leq V_{max,q}$, $q = 1, \cdots, N$), and maximum allowable currents for the n_l power lines (i.e., $I_l \leq I_{max}$, $l = 1, \cdots, n_l$).

The objective function could take into account both technical and economic aspects, and it is typically expressed as a weighted sum of O normalized design objectives:

$$J(y, \Gamma) = \alpha_{F_1} \frac{F_1(y, \Gamma)}{\bar{F}_1} + \alpha_{F_2} \frac{F_2(y, \Gamma)}{\bar{F}_2} + \cdots + \alpha_{F_O} \frac{F_O(y, \Gamma)}{\bar{F}_O} \tag{4}$$

The typical design objectives that should be minimized are:

- the active power losses:

$$F_1 = P_g - P_l \geq 0 \tag{5}$$

where P_g and P_l are the total active power generated and absorbed on the network;
- the average voltage deviation:

$$F_2 = \frac{\sum_{i=1}^{n} \| V_q - V_q^\star \|}{N} \tag{6}$$

where V_q and $V_q^\star$ are the current and the desired voltage at the node q respectively, and N is the number of nodes;
- the maximum voltage deviation:

$$F_3 = \max_i \left(\| V_q - V_q^\star \| \right) = \| V_q - V_q^\star \|_\infty \tag{7}$$

Since the design objectives are competing, the voltage control problem has no unique solution and a suitable trade-off between objectives must be identified.

3. A Decentralized Solution of the Optimal Voltage Regulation Problem

To address the voltage regulation problem, an innovative solution based on a decentralized architecture is discussed here. This architecture is based on a network of N cooperative smart controllers, each regulating the voltage magnitude of a specific bus (called node too).

In this operating scenario, each controller device is equipped with three basic components:

1. a set of sensors measuring the available set of local electrical variables (i.e., voltage magnitude, active and reactive bus power);
2. a dynamical system, whose state is initialized by sensor measurements and evolves interactively with the states of nearby controllers, according to a properly designed distributed control strategy;
3. a communication interface, carrying the interaction among controllers by transmitting the state of the dynamical system and receiving the state transmitted by the other nodes.

The idea is to leverage the theoretical framework of multi-agent dynamical systems [39] to design and implement distributed cooperative control strategies allowing the optimal energy management of the whole power grid. To this aim, the N cooperative controller devices are modeled as a one-dimensional network of dynamical agents, in which each agent uses only its neighboring information to locally control the voltage magnitude of the bus, while aiming to achieve certain global coordination with all other agents. This mathematical framework is represented in Figure 1 as the composition of the following main interrelated components: (a) agent dynamics that describe the dynamics of each bus controller device; (b) communication topology, which indicates how and if a controller device obtains information about other agents, depending on the presence/absence of connecting lines; (c) distributed control action, which is implemented at the single-controller device level, and depends on both the state variables of the bus controller itself, and on information received from neighboring controller devices through the communication topology. Use of this paradigm allows the voltage controllers to assess, in a totally decentralized way, many important variables characterizing the actual operation of the grid. Thanks to this feature, each controller knows both the variables characterizing the monitored bus (sensed by in-built sensors) and the global variables describing the actual performance of the entire power grid (assessed by checking the state of the dynamical system). This allows each controller to (i) assess the evolution of the objective function describing the voltage regulation objectives and (ii) identify the proper control actions aimed at minimizing this function.

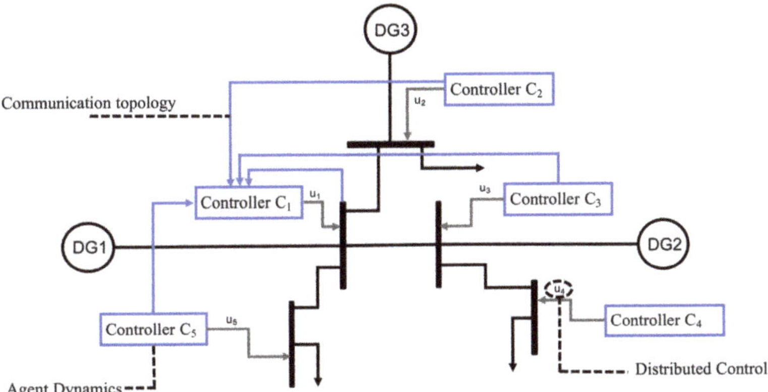

Figure 1. Cooperative smart controller network as a multi-agent dynamic system.

From this perspective, a power grid made of N_c capacitor banks, N_g generators and $N = N_g + N_c$ buses can be managed by N cooperative smart controllers. Specifically, if the i-th smart controller is associated with the i-th generation bus ($i = 1, \cdots, N_g$) of the grid, then it aims at regulating the i-th bus voltage magnitude so to achieve the desired voltage V_i^*. Conversely, if the j-th smart controller is associated with the j-th capacitor bank bus ($j = 1, \cdots, N_c$), then it aims at controlling, via distributed cooperative algorithm, the j-th bus reactive power generation capability in order to: (*i*) guarantee that its voltage magnitude achieves the desired optimal voltage value (according to (6)), as imposed by the N_g generators within the electric grid; (*ii*) reduce power losses (according to (5)).

3.1. Agent Dynamics

Within our theoretical framework, each smart device j ($j = 1, \cdots, N_c$) for the j-th capacitor bank bus of the power grid is described by the following dynamical system:

$$\dot{Q}_j(t) = u_j(t), \tag{8}$$

where $Q_j(t)$ [*p.u.*] represents the reactive power of the j-th capacitor bank bus; $u_j(t)$ is the cooperatively distributed control input that drives the reactive power, and hence the voltage magnitude, of the electrical node. It is evaluated by exploiting both local electrical measurements and electrical networks information.

Conversely, we assume that each smart device i ($i = 1, \cdots, N_g$) for the i-th generation bus within the electrical grid is described by the following dynamical system:

$$\dot{V}_i(t) = u_i(t), \tag{9}$$

where $V_i(t)$ [*p.u.*] represents the voltage magnitude of the i-th generator; $u_i(t)$ is the control input that drives the voltage magnitude of the electrical node so the achieve the desired voltage V_i^*. Please note that the i-th generator acts as a leader for the whole SG by imposing the reference voltage magnitude $V_i(t)$ for the capacitor bank buses.

3.2. Communication Topology

The communication topology indicates how and if a smart controller q ($q = 1, \cdots, N$, with $N = N_g + N_c$) obtains information about the other smart devices p ($p = 1, \cdots, N$, with $q \neq p$).

The connections among the N_c cooperative smart controller for the capacitor bank buses can be modeled as a directed graph (digraph) $\mathcal{G}_{N_c} = (\mathcal{V}, \mathcal{E}, \mathcal{A})$ of order N_c characterized by the set of nodes $\mathcal{V} = \{1, \ldots, N_c\}$ and the set of edges $\mathcal{E} \subseteq \mathcal{V} \times \mathcal{V}$. The topology of the graph is associated with an adjacency matrix with non-negative elements $\mathcal{A} = [\alpha_{j\rho}]_{N \times N}$, being $\rho = 1, \cdots, N_c$. In what follows, we assume $\alpha_{j\rho} = 1$ in the presence of a communication link from the smart device j to device ρ, otherwise $\alpha_{j\rho} = 0$. Moreover, $\alpha_{jj} = 0$ (i.e., self-edges (j, j) are not allowed). The presence of edge $(j, \rho) \in \mathcal{E}$ means that device j can obtain information from the device ρ, but not necessarily viceversa.

The presence/absence of connections among the N_c cooperative smart controller and the N_g smart controller for the generation buses is instead described by the matrix $\mathcal{A}_1 = [\alpha_{ji}]_{N_c \times N_g}$, whose elements $\alpha_{ji} = 1$ in the presence of a communication link among the smart device j and the device i, otherwise $\alpha_{ji} = 0$.

Finally, we highlight that in our application we assume that the pairs (j, ρ) and (j, i) can communicate if there exist a power transmission line among them.

3.3. Control Design

The voltage regulation control problem for the power grid can be solved by achieving two control objectives, namely:

1. designing the control strategy $u_i(t)$ in (9), leveraging local electric information, to opportunely manage the voltage magnitude of the bus i so to reach and maintain the desired reference voltage value V_i^*, i.e.,

$$\lim_{t\to\infty} \|(V_i(t) - V_i^*)\| = 0 \tag{10}$$

$\forall i = 1, \cdots, N_g$, being $V_i(t)$ the voltage magnitude of the i-th electrical node;

2. designing the distributed control input $u_j(t)$ in (8), leveraging both local and neighboring electrical information, to opportunely manage the reactive power of the bus j, hence updating its voltage magnitude $V_j(t)$ until it converges to the common reference behavior imposed by generators within the SG, i.e.,

$$\lim_{t\to\infty} \| \sum_{\rho=1}^{N_c} \alpha_{j\rho}(V_j(t) - V_\rho(t))\| \to 0$$
$$\lim_{t\to\infty} \| \sum_{i=1}^{N_g} \alpha_{ji}(V_j(t) - V_i(t))\| \to 0 \tag{11}$$

being $V_i(t)$ the voltage magnitude of the i-th generation bus and $V_\rho(t)$ the voltage magnitude of the neighboring nodes ρ ($\forall \rho = 1, \cdots, N_c$, with $j \neq \rho$).

To attain the control goal in (10), we consider, according to the literature (see e.g., [40]), for each electric node i the following Proportional controller:

$$u_i(t) = k_i(V_i(t) - V_i^*)) \tag{12}$$

where k_i is the proportional gain to be properly tuned according to the maximum admissible voltage variations for the i-th node.

Conversely, to attain the control objective in (11), we propose, for each electric node j the following consensus-based control protocol that updates its action based on the errors among the electrical state information:

$$u_j(t) = k_j \sum_{\rho=1}^{N_c} \alpha_{j\rho}(V_j(t) - V_\rho(t)) + b_j \sum_{i=1}^{N_g} \alpha_{ji}(V_j(t) - V_i(t)), \tag{13}$$

where $\alpha_{j\rho}$ models the presence/absence of communication link among the bus j and the bus ρ; α_{ji} models the presence/absence of communication link among the bus j and the generator bus i; k_j and b_j are the control gains to be tuned so to guarantee that the reactive power of the bus j does not exceed a pre-fixed operating range $[Q_{j,min}; Q_{j,max}]$.

Finally, we highlight that the exploitation of the distributed cooperative control input (13) guarantees that all the N_c buses within the SG closely converge to the common behavior imposed by the N_g generators, hence guaranteeing the optimal management of the electrical grid.

4. Case Study

To validate the effectiveness of the approach proposed in Section 3, we consider here the voltage regulation problem for the IEEE 30-bus test system depicted in Figure 2. The power grid is made of $N_g = 6$ generators (i.e., node 1, 2, 5, 8, 11, 13), $N_c = 24$ capacitor banks, $N = 30$ buses, and $n_l = 41$ lines. Load information, lines impedance, as well as reactive power limits, are reported in [41]. Numerical analysis has been carried out by exploiting the MATLAB© platform.

The initial condition for the N cooperative agents within the grid as well as the selected control gains for both the control protocol in (12) and the one in (13) are listed in Table 1. Finally, the desired voltage value V_i^* for the generation buses N_g are selected as follows: $[V_1^*, V_2^*, V_5^*, V_8^*, V_{11}^*, V_{13}^*] = [1.05, 1.02, 1.05, 1.03, 1.05, 1.02] \ [p.u.]$. Our aim is to show how the proposed solution can ensure in a totally distributed fashion the desired optimal voltage magnitude of the whole grid with reduced power losses.

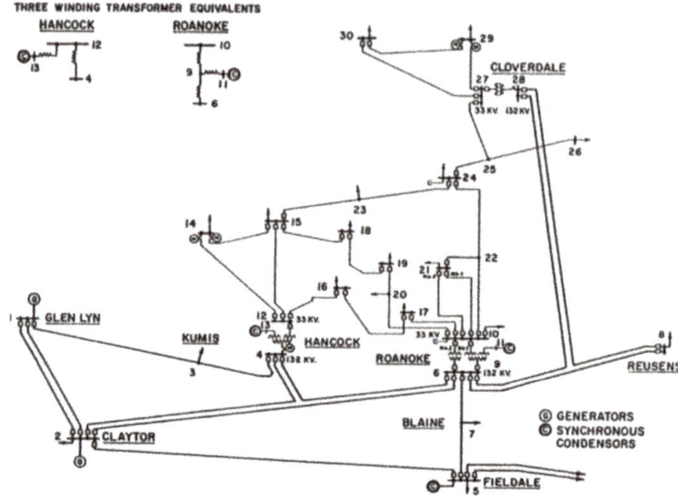

Figure 2. The IEEE 30-bus test system.

Table 1. Simulation parameters for the IEEE 30-bus test system.

	Initial Conditions
Voltage magnitude of generation bus i [$p.u.$]	$V_1(0) = 1.02; V_2(0) = 1.01; V_5(0) = 1.03;$ $V_8(0) = 1.04; V_{11}(0) = 1.01; V_{13}(0) = 1.03$
Reactive power of capacitor bank bus j [$p.u.$]	$Q_3(0) = -0.012; Q_4(0) = -0.016;$ $Q_6(0) = -0.005; Q_7(0) = -0.109;$ $Q_9(0) = -0.005; Q_{10}(0) = -0.02$ $Q_{12}(0) = -0.075; Q_{14}(0) = -0.016;$ $Q_{15}(0) = -0.025; Q_{16}(0) = -0.018;$ $Q_{17}(0) = -0.058; Q_{18}(0) = -0.009;$ $Q_{19}(0) = -0.034; Q_{20}(0) = -0.007;$ $Q_{21}(0) = -0.112; Q_{22}(0) = -0.005;$ $Q_{23}(0) = -0.016; Q_{24}(0) = -0.067;$ $Q_{25}(0) = -0.005; Q_{26}(0) = -0.023;$ $Q_{27}(0) = -0.005; Q_{28}(0) = -0.005;$ $Q_{29}(0) = -0.009; Q_{30}(0) = -0.019.$
	Control Gains
Control gains k_i Control gains k_j Control gains b_j	$k_i = 5 \quad i = 1, 2, 5, 8, 11, 13$ $k_j = 10 \quad \forall j \in N_c$ $b_j = 15 \quad \forall j \in N_c$

Results in Figures 3–5 show the effectiveness of the proposed control approach in ensuring the control goal in (10) and (11). Indeed, as depicted in Figure 3, thanks to the control action (12), the smart controllers for the generation buses ensure that the corresponding voltage magnitude converge to the desired values V_i^* in 1 [s]. Accordingly, the smart controller device j ($j \in N_c$) dynamically acts on the reactive power generation capability of the j-th electrical node (i.e., it produces or absorbs the bus reactive power; see Figure 5) and due to cooperative control action in (13) guarantees that its voltage magnitude converges in 1 [s] to the desired optimal value [1.02; 1.05] [$p.u.$] as imposed by the N_g generators of the electrical grid. Indeed, the mean grid voltage of the N_c electrical nodes, i.e., $V_{mean} = 1.03$ [$p.u.$] is equal to the mean voltage imposed by the N_g generators. This confirms the benefit of the proposed decentralized approach in ensuring the electrical grid synchronization to the reference behavior imposed by generators only leveraging local and neighboring information and without the need for knowing global information about the whole grid. The proposed approach hence provides a distributed control solution allowing the power grid to be self-organized. The benefits of the approach are the scalability of the solution, which makes the grid to be easily re-configurable, and the low computational burden required for the controlling purposes.

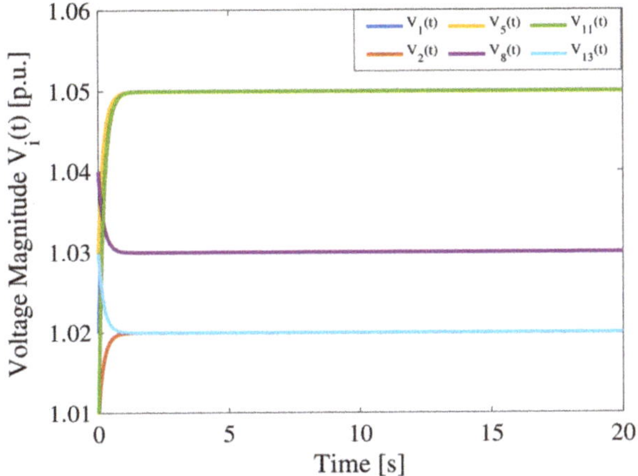

Figure 3. Time history of the voltage magnitude $V_i(t)$ $[p.u.]$ for $i \in N_g$.

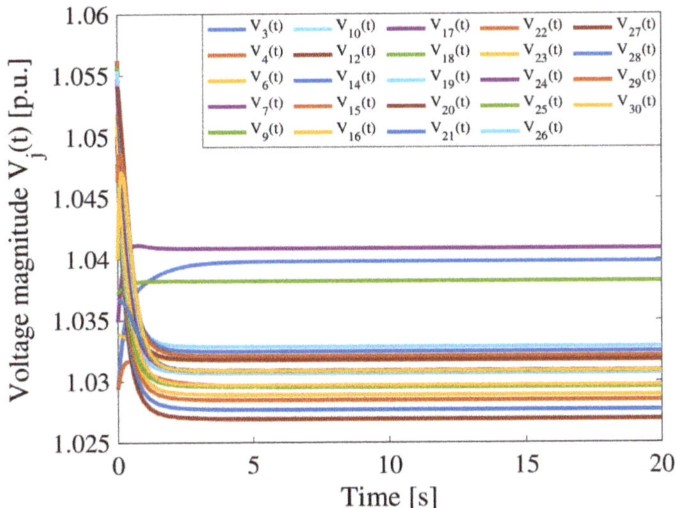

Figure 4. Time history of the voltage magnitude $V_j(t)$ $[p.u.]$ for $j \in N_c$.

Finally, we remark that due to the scalability features of the proposed approach, the convergence settling time still remains equal to 1 $[s]$ when increasing the number of devices, differently from the centralized solution where this increasing number may significantly affects the dynamic performance of the whole grid.

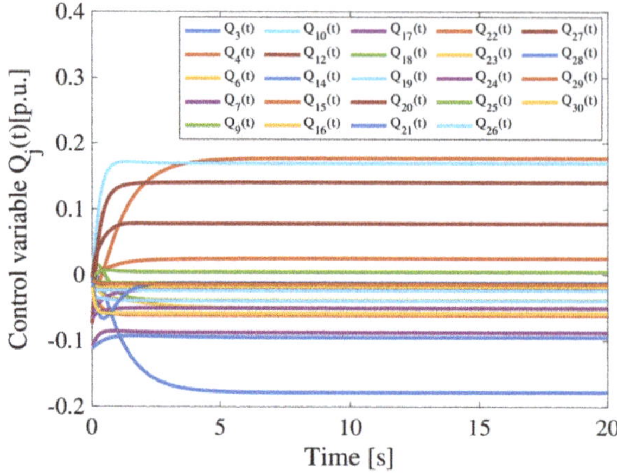

Figure 5. Time history of the reactive power $Q_j(t)$ [$p.u.$] for $j \in N_c$.

Robustness to Variable Load

Here we show the robustness of the proposed approach by considering percentage variations of all the loads compared to the nominal values reported in [41]. Specifically, as illustrative example, we take into account that load, indicated with $L(t)$, varies over time according to Figure 6 (i.e., maximum variations of $\pm 50\%$). Results in Figures 7 and 8 disclose that despite the presence of load variations acting on the whole grid, the proposed approach promptly reacts to load changes by recovering the desired optimal voltage magnitude as imposed by the N_g generators. Specifically, in the time interval $5 \leq t < 10$ it is possible to appreciate that after increasing of the 30% of all the loads within the grid (see Figure 6), the smart controllers dynamically regulate the voltage magnitude of the j-th bus via production or absorption of reactive power. The same behavior also occurs when there exist both a variation of 50% in the time interval $10 \leq t < 15$ and a variation of -50% in the time interval $15 \leq t < 20$.

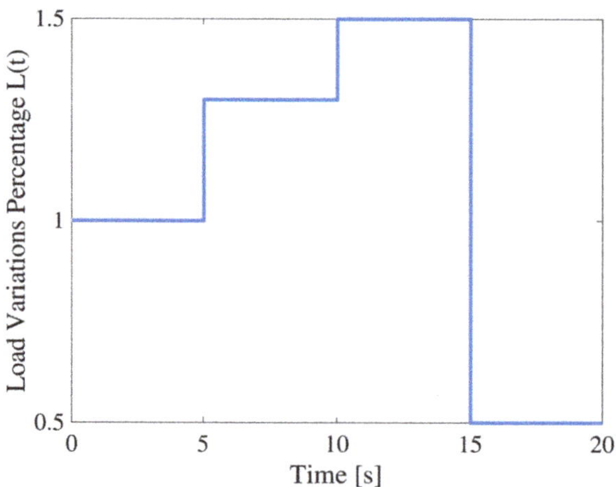

Figure 6. Load Variations Percentage L(t).

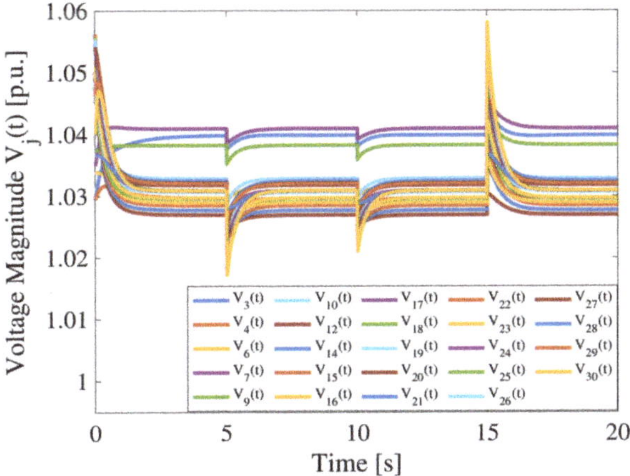

Figure 7. Robustness with respect to variable load: time history of the voltage magnitude $V_j(t)$ $[p.u.]$ for $j \in N_c$.

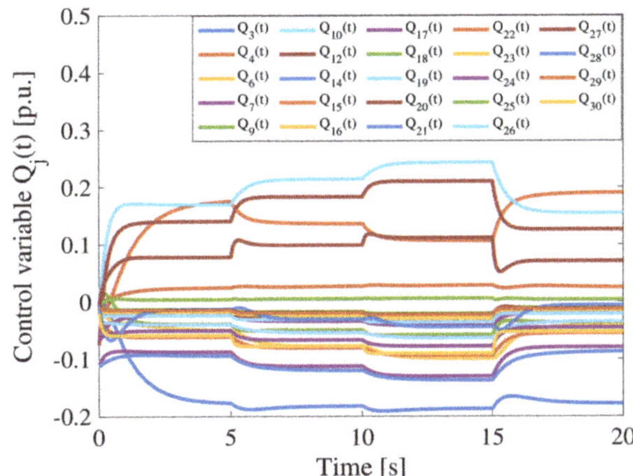

Figure 8. Robustness with respect to variable load: time history of the reactive power $Q_j(t)$ $[p.u.]$ for $j \in N_c$.

5. Conclusions

MAS-based architectures are considered as the most promising enabling methodology for decentralized voltage regulation in modern power distribution systems, where the massive pervasion of renewable power generators could hinder the deployment of hierarchical and centralized control paradigms. Although the conceptualization of MAS-based solutions has been widely explored in the literature, their deployment in realistic operation scenario is still at its infancy, and several open problems need to be addressed in order to identify the most effective computing paradigm, which reliably solves the voltage regulation problem, exhibiting high resilience to internal and external perturbations, high scalability to support an exponential growth of distributed energy resources, and low computational requirements.

In the light of these needs, this paper proposed a novel decentralized control architecture for the online voltage regulation in active power grids, which is based on a network of cooperative dynamic agents equipped with consensus protocols, and interacting with each other through wireless/wired communication networks. Thanks to the adoption of this cooperative paradigm, each agent regulates the voltage magnitude of a specific generation/load bus so to achieve a synchronization behavior to desired voltage magnitude without the need for fusion data center. This feature makes the proposed solution self-organized, decentralized, and scalable, hence resulting a promising alternative to solve the voltage regulation problem in modern smart grids.

Simulation results, carried out for the realistic case of study of the IEEE 30-bus test system, confirm the effectiveness and the robustness of the approach in ensuring that all the electrical buses reach and maintain the desired synchronization behavior.

Author Contributions: Conceptualization, A.A., A.P., S.S., A.V. and D.V.; Methodology, A.A., A.P., S.S. and A.V.; Software, A.A., A.P., S.S. and A.V.; writing–original draft preparation, A.A., A.P., S.S. and A.V.; writing–review and editing, A.A., A.P., S.S., A.V. and D.V.

Funding: This research received no external funding.

Conflicts of Interest: The authors declare no conflict of interest.

References

1. Cossent, R.; Gómez, T.; Olmos, L. Large-scale integration of renewable and distributed generation of electricity in Spain: Current situation and future needs. *Energy Policy* **2011**, *39*, 8078–8087. [CrossRef]
2. Liu, Y.; Bebic, J.; Kroposki, B.; De Bedout, J.; Ren, W. Distribution system voltage performance analysis for high-penetration PV. In Proceedings of the 2008 IEEE Energy 2030 Conference, Atlanta, GA, USA, 17–18 November 2008; pp. 1–8.
3. Rahman, M.M.; Arefi, A.; Shafiullah, G.; Hettiwatte, S. A new approach to voltage management in unbalanced low voltage networks using demand response and OLTC considering consumer preference. *Int. J. Electr. Power Energy Syst.* **2018**, *99*, 11–27. [CrossRef]
4. Walling, R.; Saint, R.; Dugan, R.C.; Burke, J.; Kojovic, L.A. Summary of distributed resources impact on power delivery systems. *IEEE Trans. Power Deliv.* **2008**, *23*, 1636–1644. [CrossRef]
5. Jamal, T.; Urmee, T.; Calais, M.; Shafiullah, G.; Carter, C. Technical challenges of PV deployment into remote Australian electricity networks: A review. *Renew. Sustain. Energy Rev.* **2017**, *77*, 1309–1325. [CrossRef]
6. Kenneth, A.P.; Folly, K. Voltage rise issue with high penetration of grid connected PV. *IFAC Proc. Vol.* **2014**, *47*, 4959–4966. [CrossRef]
7. Karimi, M.; Mokhlis, H.; Naidu, K.; Uddin, S.; Bakar, A. Photovoltaic penetration issues and impacts in distribution network—A review. *Renew. Sustain. Energy Rev.* **2016**, *53*, 594–605. [CrossRef]
8. Stetz, T.; Marten, F.; Braun, M. Improved low voltage grid-integration of photovoltaic systems in Germany. *IEEE Trans. Sustain. Energy* **2013**, *4*, 534–542. [CrossRef]
9. Varma, R.K.; Rangarajan, S.S.; Axente, I.; Sharma, V. Novel application of a PV solar plant as STATCOM during night and day in a distribution utility network. In Proceedings of the 2011 IEEE/PES Power Systems Conference and Exposition, Phoenix, AZ, USA, 20–23 March 2011; pp. 1–8.
10. Caldon, R.; Coppo, M.; Turri, R. Distributed voltage control strategy for LV networks with inverter-interfaced generators. *Electr. Power Syst. Res.* **2014**, *107*, 85–92. [CrossRef]
11. Molina-García, Á.; Mastromauro, R.A.; García-Sánchez, T.; Pugliese, S.; Liserre, M.; Stasi, S. Reactive power flow control for PV inverters voltage support in LV distribution networks. *IEEE Trans. Smart Grid* **2017**, *8*, 447–456. [CrossRef]
12. Perera, B.K.; Ciufo, P.; Perera, S. Point of common coupling (PCC) voltage control of a grid-connected solar photovoltaic (PV) system. In Proceedings of the IECON 2013 39th Annual Conference of the IEEE Industrial Electronics Society, Vienna, Austria, 10–13 November 2013; pp. 7475–7480.
13. Raju, L.; Milton, R.; Mahadevan, S. Multi agent systems based distributed control and automation of micro-grid using MACSimJX. In Proceedings of the 2016 10th International Conference on Intelligent Systems and Control (ISCO), Coimbatore, India, 7–8 January 2016; pp. 1–6.
14. Fallahzadeh-Abarghouei, H.; Nayeripour, M.; Waffenschmidt, E.; Hasanvand, S. A new decentralized voltage control method of smart grid via distributed generations. In Proceedings of the 2016 International Energy and Sustainability Conference (IESC), Cologne, Germany, 30 June–1 July 2016; pp. 1–6.
15. Higgins, N.; Vyatkin, V.; Nair, N.; Schwarz, K. Intelligent decentralised power distribution automation with IEC 61850, IEC 61499 and holonic control. *IEEE Trans. Syst. Mach. Cybern. C* **2010**, *40*, 1–12.

16. Demirok, E.; Gonzalez, P.C.; Frederiksen, K.H.; Sera, D.; Rodriguez, P.; Teodorescu, R. Local reactive power control methods for overvoltage prevention of distributed solar inverters in low-voltage grids. *IEEE J. Photovolt.* **2011**, *1*, 174–182. [CrossRef]

17. Turitsyn, K.; Sulc, P.; Backhaus, S.; Chertkov, M. Local control of reactive power by distributed photovoltaic generators. In Proceedings of the 2010 First IEEE International Conference on Smart Grid Communications, Gaithersburg, MD, USA, 4–6 October 2010; pp. 79–84.

18. Demirok, E.; Sera, D.; Rodriguez, P.; Teodorescu, R. Enhanced local grid voltage support method for high penetration of distributed generators. In Proceedings of the IECON 2011 37th Annual Conference of the IEEE Industrial Electronics Society, Hungary, Budapest, 26–30 August 2011; pp. 2481–2485.

19. Martí, P.; Velasco, M.; Fuertes, J.M.; Camacho, A.; Miret, J.; Castilla, M. Distributed reactive power control methods to avoid voltage rise in grid-connected photovoltaic power generation systems. In Proceedings of the 2013 IEEE International Symposium on Industrial Electronics, Taipei, Taiwan, 28–31 May 2013; pp. 1–6.

20. Tanaka, K.; Oshiro, M.; Toma, S.; Yona, A.; Senjyu, T.; Funabashi, T.; Kim, C.H. Decentralised control of voltage in distribution systems by distributed generators. *IET Gen. Transm. Distrib.* **2010**, *4*, 1251–1260. [CrossRef]

21. Toma, S.; Senjyu, T.; Miyazato, Y.; Yona, A.; Funabashi, T.; Saber, A.Y.; Kim, C.H. Optimal coordinated voltage control in distribution system. In Proceedings of the 2008 IEEE Power and Energy Society General Meeting-Conversion and Delivery of Electrical Energy in the 21st Century, Pittsburgh, PA, USA, 20–24 July 2008; pp. 1–7.

22. Shalwala, R.; Bleijs, J. Voltage control scheme using Fuzzy Logic for residential area networks with PV generators in Saudi Arabia. In Proceedings of the 2010 Joint International Conference on Power Electronics, Drives and Energy Systems & 2010 Power India, New Delhi, India, 20–23 December 2010; pp. 1–6.

23. Gupta, N.; Garg, R.; Kumar, P. Asymmetrical Fuzzy logic control to PV Module Connected micro-grid. In Proceedings of the 2015 Annual IEEE India Conference (INDICON), New Delhi, India, 17–19 December 2015; pp. 1–6.

24. Loia, V.; Vaccaro, A. A decentralized architecture for voltage regulation in smart grids. In Proceedings of the 2011 IEEE International Symposium on Industrial Electronics, Gdansk, Poland, 27–30 June 2011; pp. 1679–1684.

25. Loia, V.; Vaccaro, A.; Vaisakh, K. A self-organizing architecture based on cooperative fuzzy agents for smart grid voltage control. *IEEE Trans. Ind. Inf.* **2013**, *9*, 1415–1422. [CrossRef]

26. Hojo, M.; Hatano, H.; Fuwa, Y. Voltage rise suppression by reactive power control with cooperating photovoltaic generation systems. In *Electricity Distribution-Part*; IET: Stevenage, UK, 2009; Volume 2.

27. Wang, X.; Wang, C.; Xu, T.; Guo, L.; Fan, S.; Wei, Z. Decentralised voltage control with built-in incentives for participants in distribution networks. *IET Gen. Transm. Distrib.* **2018**, *12*, 790–797. [CrossRef]

28. Zhang, Z.; Ochoa, L.F.; Valverde, G. A novel voltage sensitivity approach for the decentralized control of DG plants. *IEEE Trans. Power Syst.* **2018**, *33*, 1566–1576. [CrossRef]

29. Liu, H.J.; Shi, W.; Zhu, H. Decentralized dynamic optimization for power network voltage control. *IEEE Trans. Signal Inf. Proc. Netw.* **2017**, *3*, 568–579. [CrossRef]

30. Lin, W.; Bitar, E. Decentralized stochastic control of distributed energy resources. *IEEE Trans. Power Syst.* **2018**, *33*, 888–900. [CrossRef]

31. Antoniadou-Plytaria, K.E.; Kouveliotis-Lysikatos, I.N.; Georgilakis, P.S.; Hatziargyriou, N.D. Distributed and decentralized voltage control of smart distribution networks: Models, methods, and future research *IEEE Trans. Smart Grid* **2017**, *8*, 2999–3008. [CrossRef]

32. Petrillo, A.; Salvi, A.; Santini, S.; Valente, A.S. Adaptive synchronization of linear multi-agent systems with time-varying multiple delays. *J. Frankl. Inst.* **2017**, *354*, 8586–8605. [CrossRef]

33. Petrillo, A.; Pescapé, A.; Santini, S. A collaborative approach for improving the security of vehicular scenarios: The case of platooning. *Comput. Commun.* **2018**, *122*, 59–75. [CrossRef]

34. Petrillo, A.; Salvi, A.; Santini, S.; Valente, A.S. Adaptive multi-agents synchronization for collaborative driving of autonomous vehicles with multiple communication delays. *Trans. Res. C Emerg. Technol.* **2018**, *86*, 372–392. [CrossRef]

35. Di Vaio, M.; Fiengo, G.; Petrillo, A.; Salvi, A.; Santini, S.; Tufo, M. Cooperative shock waves mitigation in mixed traffic flow environment. *IEEE Trans. Intell. Trans. Syst.* **2019**, *99*, 1–15. [CrossRef]

36. Dorfler, F.; Bullo, F. Synchronization and transient stability in power networks and nonuniform Kuramoto oscillators. *SIAM J. Control Optim.* **2012**, *50*, 1616–1642. [CrossRef]

37. Zheng, J.; Jamalipour, A. *Wireless Sensor Networks: A Networking Perspective*; John Wiley & Sons: Hoboken, NJ, USA, 2009.

38. Loia, V.; Furno, D.; Vaccaro, A. Decentralised smart grids monitoring by swarm-based semantic sensor data analysis. *Int. J. Syst. Control Commun.* **2013**, *5*, 1–14. [CrossRef]

39. Olfati-Saber, R.; Fax, J.A.; Murray, R.M. Consensus and cooperation in networked multi-agent systems. *Proc. IEEE* **2007**, *95*, 215–233. [CrossRef]

40. He, D.; Shi, D.; Sharma, R. Consensus-based distributed cooperative control for microgrid voltage regulation and reactive power sharing. In Proceedings of the IEEE PES Innovative Smart Grid Technologies, Istanbul, Turkey, 12–15 October 2014; pp. 1–6.

41. Shahidehpour, M.; Wang, Y. *Communication and Control in Electric Power Systems: Applications of Parallel and Distributed Processing*; John Wiley & Sons: Hoboken, NJ, USA, 2004.

MDPI

St. Alban-Anlage 66

4052 Basel

Switzerland

Tel. +41 61 683 77 34

Fax +41 61 302 89 18

www.mdpi.com

Energies Editorial Office

E-mail: energies@mdpi.com

www.mdpi.com/journal/energies

www.ingramcontent.com/pod-product-compliance
Lightning Source LLC
LaVergne TN
LVHW070626170726
843515LV00004B/183